MW00710009

CASE

Motorbooks International

FARM TRACTOR COLOR HISTORY

CASE GP TRACTORS

Robert N. Pripps & Andrew Morland

For my newest grandson, Noah Adam Pripps

First published in 1996 by Motorbooks International Publishers & Wholesalers, 729 Prospect Avenue, PO Box 1, Osceola, WI 54020-0001 USA

Library of Congress Cataloging-in-Publication Data

Pripps, Robert N.
Case GP tractors/Robert N. Pripps, Andrew Morland.
p. cm.—(Motorbooks International farm tractor color history)
Includes index.
ISBN 0-7603-0116-6 (pbk.: alk. paper)
1. Case tractors—History. 2. Case tractors—Pictorial works. I. Morland, Andrew.
II. Title. III. Series.
TL233.6.C37P75 1996
629.225—dc20 96-22154

On the front cover: A 1944 Case Model SC pulling a Case Centennial 2x16 plow.

On the frontispiece: New styling changes marked the 1957 Case tractor line. The new styling was applied to the entire Case tractor sizes line, with the headlights high in the grille and a squared-off appearance that was both practical and purposeful.

On the title page: A 1950 Model SC and a Case thresher. The tractor is owned by Kelly Clevinger.

On the back cover: Top: A 1929 Case Model L. The Model L used a roller chain final drive, three-speed transmission, and a rear power take-off. Bottom: This Model 830 HC is a high-clearance version equipped with Case-O-Matic drive. It is a 1964 model and is owned by J.R. Gyger.

Printed in Hong Kong

Contents

	Acknowledgments	6
	Preface	7
	Introduction	9
Chapter 1	Heritage	13
Chapter 2	Early Years	21
Chapter 3	The Crossmotor Era, 1916–28	29
Chapter 4	General Purpose Tractors, 1929–39	43
Chapter 5	The Flambeau Red Era, 1939–55	65
Chapter 6	Tractors of the Desert Sunset, 1956–76	93
Appendix	Clubs, Shows, and Literature	126
	Index	128

Acknowledgments

Photographer Andrew Morland and I offer our special thanks to the dedicated Case collectors whose places we visited and whose tractors we photographed in 1994: Weston Rink, John Davis, Jay Foxworthy, Elwood and Don Voss, Clyde Barrows, J.R. and Jay Gyger, and John Thierer. Also in 1994, we attended the Case exposition at the Waukee (Iowa) show. There we photographed outstanding examples of restored Case tractors owned by Loren Simmons, Tom Graverson, Dick Nesselroad, Loren Engle, Warren Kemper, Dan Buckert, Kelly Clevenger, and Scott Fuller; our thanks to them, as well.

Special thanks goes to Mr. Dave Rogers of the Case Corporation in Racine, Wisconsin, for all the archive pictures and other help during the preparation of this book. I once thought that Deere & Company stood alone in their concern for heritage and nostalgia, but Case has proven to be their equal.

Thanks to David Erb for fact-checking and working on captions. Without his help, this book wouldn't be what it is.

Our thanks, as well, to the staff at Motorbooks International for their usual fine job of design, editing, and marketing.

—Robert N. Pripps

Preface

Are not all Case tractors usable for purposes in general? What is the definition of a general purpose tractor? What is the difference between a general purpose tractor and a utility tractor?

The first Case tractor to be officially called "general purpose" was the Model CC of 1930. By then, many other manufacturers were also offering general purpose machines.

The term "general purpose," or "all-purpose," really originated with the Moline Universal of 1914. The Universal could be used for various plowing, tilling, and belt jobs. The concept really took root, however, with the 1924 Farmall. The Farmall could do the plowing, tilling, and belt jobs, plus drive implements (such as the new shaft-powered binders) via the rear power take-off (PTO).

When the John Deere Models A and B were introduced in 1934 and 1935, they sported "GP" (General Purpose) initials on their hood. By then, there were conventionally three kinds of farm tractors: GP, Standard, and Orchard. After the introduction of the Ford-Ferguson in 1939, the utility configuration began to replace all three.

Although this book is mainly about Case's general purpose tractors, all models are covered because they are closely related. After 1955, all Case tractors fit the "General Purpose" definition.

The definition of general purpose (as it has evolved): A tractor that an owner can use for plowing, tilling, cultivating, and power take-off work; one that has adequate crop clearance, adjustable wheel tread, individual steering brakes, and a power implement lift.

Introduction

There are no more loyal tractor fans than Case tractor fans. This most likely stems from the character of the company that produced these tractors—which reflects the character of the company founder, J.I. Case.

When Jerome Increase Case started in business in 1843, he was one of several in the agricultural field. John Deere, Cyrus McCormick, Daniel Massey, James Oliver, and later, Henry Ford and Harry Ferguson were others. These men had a profound impact not only on agriculture, but also on society in general—they changed people's daily lives. Each of these men had a rare combination of inventive genius, perseverance, business sense, and the charisma to attract the talents they lacked.

Jerome Increase Case was born in Oswego, New York, on December 11, 1819 (some say 1818). He was the son of Caleb Case, a farmer, and Deborah Jackson Case, a member of the family that produced President Andrew Jackson. Interestingly, Andrew Jackson would, in 1836, sign the bill creating the Wisconsin territory, the future home of the Case Corporation.

Case became interested in farm power at a young age. An article in the *Genessee Farmer* announcing a demonstration of the Groundhog thresher caught 16-year-old Jerome's eye. He badgered his father to see it. Caleb Case was suitably impressed and bought one of the first Groundhog threshing machines in the territory. In fact, Caleb Case became an agent for the Pitts Brothers, who made the Groundhog. Young Jerome assisted his father in operating the machines and did custom threshing for the next five seasons. The long hours of threshing gave him ample time to consider the machine's limitations. It also became apparent to young Jerome that the future of grain harvesting was not in the

The Case trademark for over a century was Old Abe, originally a live bald eagle mascot from Civil War days, named after President Abe Lincoln. This trademark was used by the Case Company until 1969, and again in the late 1980s. *Case Archives*

The Pope threshing machine of 1833. This was the forerunner of the Pitts Groundhog thresher that J.I. Case first used in New York. *Smithsonian*

The Crossmotor tractors were Case's mainstay in the late 1910s and 1920s. This is a 1920 Model 15-27, which was built from 1919 to 1924. Owner: John Davis, Maplewood, Ohio.

lake area of New York, but in the Upper Mississippi River Valley, and in the Great Plains of the Midwest.

The 22-year-old Case heard stories of homesteaders with vast plots of land suitable for raising wheat. These resources were said to be untapped due to a lack of available man-power. Young Case, then completing his first year of college at Rensselaer Academy, sensed his opportunity. On credit, he bought six Pitts Groundhog threshers and horsepowers (horse treadmills) from his father and started west via the new Erie Canal for a frontier town called Chicago. His final destination was Roch-ester, Wisconsin, a town named after his hometown of Rochester, New York. This was the time of the great Yankee migration from New England to the burgeoning Midwest. Some of the Case family acquaintances had already moved to Rochester, Wisconsin.

J.I. Case arrived in Chicago during the harvest of 1842. He bought a wagon and team, loaded his threshers, and started for Racine County, Wisconsin. As he traveled, he stopped at prosperous-looking farms along the way and

demonstrated the operation of the Groundhog. Thus, he sold five of the six Groundhogs he had brought with him while on his way to Rochester, a town not far from Racine. He kept one for custom threshing.

During the winter of 1842–43, Case took room and board with the family of Seth Warner. With the help of a carpenter (curiously named Stephen Thresher) who also roomed with the family, he built a threshing machine along the Groundhog line, but with improvements. Case was pleased with the performance of his machine during the harvest of 1843. With that, Case rented a shop in Racine to manufacture threshers for the next season.

Up until this point, threshing machines did only that—they threshed the heads from the stalks and deposited the whole works on the ground below the machine. J.I. Case saw the benefit of combining the fanning mill directly into the thresher. He sought out Richard Ela, a fanning mill maker from New Hampshire. With the help of Ela, Case incorporated a fan into the design of his machine, which he had ready for the harvest of 1844. This new machine would thresh and separate, depositing the straw in one location and cleaned grain in another.

To say that the Case separating thresher was a success would be an understatement. By 1847, a new plant had been constructed in Racine, complete with its own steam power plant and its own foundry. Case, not yet 30 years old, had become Racine's largest employer.

Next, Case acquired patent rights for a vibrator apparatus designed as an add-on to the Groundhog thresher, and built it into his machine. These were produced, with periodic improvements, for about 15 years along with horsepowers of Case's design. In 1869, Case introduced the Case Eclipse thresher, an apron-less machine that pioneered the use of straw racks. Also in 1869, the first Case steam engine was belted to a threshing machine. J.I. Case died in 1892, the year after the first Case internal combustion tractor rolled out for testing.

The J.I. Case Company had become a corporation in 1880, with Case as president. With able officers in place, the company continued to prosper even after the death of its founder. It went on to become a full-line producer of agricultural and industrial machinery.

In 1985, Tenneco purchased the International Harvester agricultural division and added this to the Case line. The best of both companies was the result of this buyout. Combines, for instance, remained as they had been under IH. Small tractors made in England and Europe by IH became part of the Case line, and transmission ideas and parts from IH were used in the Case Magnum tractor line. While the new equipment bears the Case-IH name, the company name remains Case.

Chapter 1

Heritage

The saying, "Success has many fathers," applies to the Case line of general purpose tractors. As the history of the J.I. Case Company unfolded, a network of acquired technology and capability was fed in. This chapter describes the roots of the J.I. Case Company, especially those relating to tractors. Like the roots of a mighty oak, these go deep into the soil of history. As with the oak, the sturdy roots helped the company weather many storms.

The thresher J.I. Case built for the 1844 season was a great success, and he began negotiating with the Rochester city fathers for the rights to build another factory millrace on the Fox River (the water power would be needed for the new shop he intended to set up to manufacture threshers). Water power rights in Rochester were closely held by a small group, and this group did not see fit to allow anyone else into its circle. Case's request was denied. The following day, J.I. Case moved lock, stock, and barrel to Racine, then a town of about 3,000 souls 25 miles east on the shores of Lake Michi-

LEFT
Cutting grain with a cradle was slow enough that primitive threshing methods were adequate for most farmers. With the advent of the reaper and binder, grain harvesting was more rapid. This put pressure on the development of better means of threshing. Not many years passed before one thresher could keep up with many reapers. *Smithsonian*

ABOVE
Case Agitator threshing machines, introduced in 1880, revolutionized the industry. Their superiority soon proved itself by increasing market share. Later refinements to the Agitator used iron bearing retainer plates on the sides, which led to the name "Ironsides Agitator." This design soon became the industry standard.

gan. He rented a building on the Root River, (which is "Racine" in French) calling his operation the Racine Threshing Machine Works. It was Rochester's loss.

By 1847, the business had prospered sufficiently that expanded quarters were needed. Case then built a three-story brick factory. By 1848, when Wisconsin became a state, he was Racine's largest employer. Another six years saw the addition of a foundry and the switch to steam rather than water power for the factory.

Meanwhile, in 1849, Case married Lydia Ann Bull, of Yorkville, Wisconsin. She was introduced to Jerome by her brother, Stephen, who would himself play an important part in Case's company. J.I. Case and Lydia had seven children, although only four lived to maturity, three girls (Henrietta, Jesse, and Amanda) and a boy (Jackson). Lydia died in 1909, outliving her husband by 18 years.

As J.I. Case's enterprise prospered, he became interested in banking. He was founder and president of Racine's Manufacturer's National Bank, as well as involved in several other banks. Case was a firm believer that manufacturers should be in the business of financing farm equipment purchases. One can only imagine the problems of building, selling, and financing these large machines in the early nineteenth century, before the establishment of such banks. Prior to this time, Case established lines of credit with local merchants. During a few bad winters, employees were paid mostly in script, which could be exchanged for merchandise. Financing was a larger problem for the farmer. Paying for the threshing machine was generally a three-season process. When a farmer failed to pay, collection was difficult. Case, like the other implement makers, established collection departments. Robert Baker was hired in 1860 to head Case's collection operation.

Along with Baker, Manessa Erskine was hired to head the mechanical department and Case's brother-in-law, Stephen Bull, became Case's personal assistant. Together with J.I. Case, these men became known as the Big Four. The business had now grown to the point where it was more than one man could handle. Therefore, a partnership was formed with the Big Four as equal partners. The name J.I. Case & Company was adopted.

During these early turbulent years, J.I. Case—a gruff and abrupt man—was also successful in politics. He served two back-to-back terms as mayor of Racine, and one as a Wisconsin state senator.

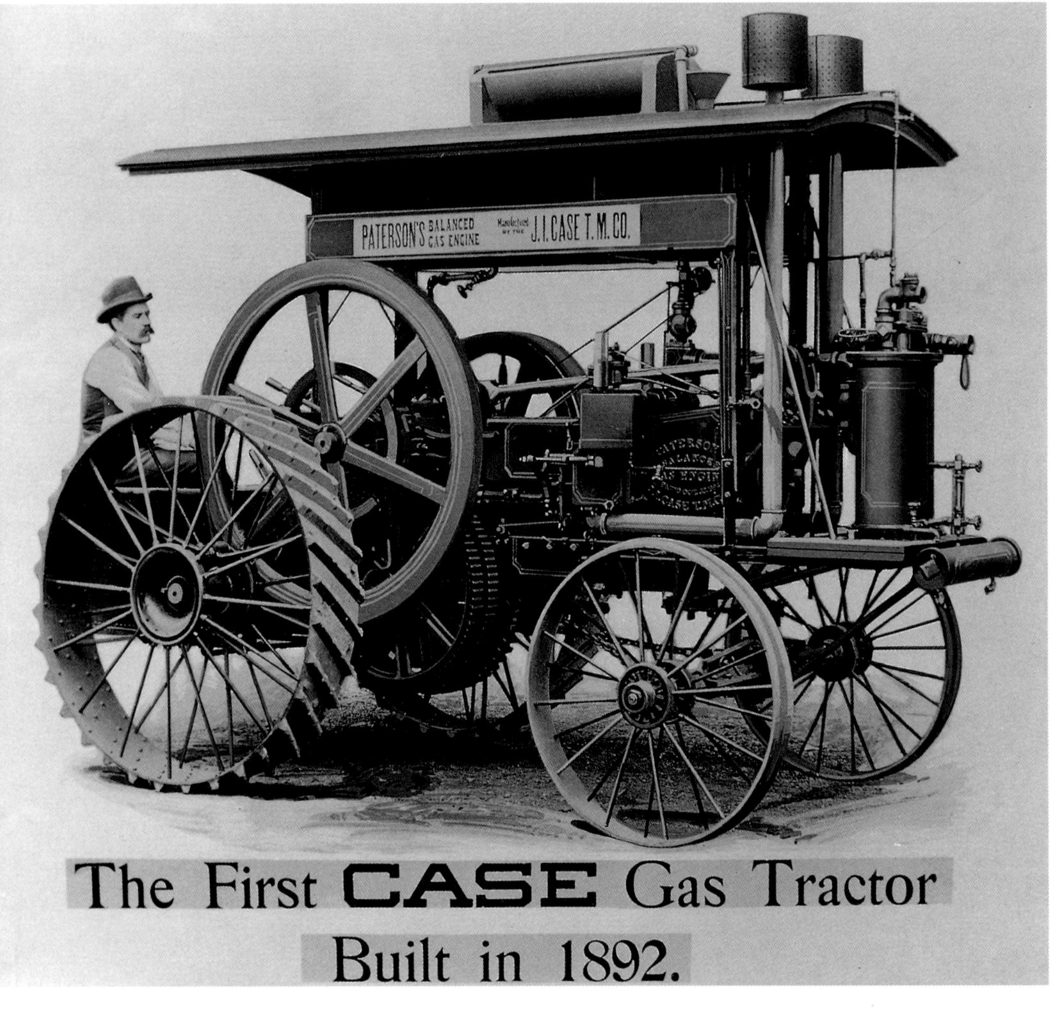

This is the only known photograph of the first Case tractor. Built by William Paterson, the unit used an unusual engine design featuring two opposed pistons connected by linkage to the same crankshaft. The unit suffered from primitive carburetion and ignition problems. *Case Archives*

Old Abe

After the end of the American Civil War, Case adopted the eagle trademark. The adoption of this symbol is shrouded in myths and fables. The logo is based on a real bird, an eagle, named Old Abe. A Chippewa Indian named Chief Sky took the bird, then an eaglet, from its nest along the Flambeau River near what is now Park Falls, Wisconsin. The Indian traded the bird for a bushel of corn to Dan McCann of Jim Falls, Wisconsin. McCann tamed the bird, keeping it tethered or in a cage made from a barrel. The McCann children scoured the vicinity for mice, rabbits, and grouse to feed the growing eagle. In 1861, McCann sold the bird for $2.50 to Company C of the Eighth Wisconsin

Regiment for a mascot, the funds having been donated by a tavern owner whose bar the men patronized. Company C named the mascot "Old Abe," for President Abraham Lincoln. One of the soldiers made Old Abe a perch where he could be carried into battle just to the left of the colors. Old Abe seemed to react to the men's cheering and their band music. He would screech or fly at the end of his tether, flapping his three-foot wings. The eagle's antics, his responsiveness to the men, and his seemingly fierce demeanor made Old Abe an ideal mascot and a real celebrity.

Case had witnessed the mustering out of Company C of the Eighth Wisconsin Regiment. He was impressed by the sight and the sound of Old Abe, who would spread his wings and give a war cry on command. Case immediately decided to make Old Abe the company symbol. At first, Old Abe was shown sitting on a branch. Later, he was shown perched upon the world. He was "The Sign of Mechanical Excellence the World Over."

As a footnote to history, the large cast iron Case Eagle that had stood in front of Case's Emerson-Brantingham plant in Rockford, Illinois, was given to the City of Park Falls, Wisconsin, when the plant closed in 1971. It was on display in the city's outdoor pavilion for many years, but recently has come up missing—flown the coop, so to speak. City fathers would like to hear from anyone knowing its whereabouts.

New Threshers

In 1869, J.I. Case & Company introduced a new type of thresher; the Eclipse. It was a departure from the traditional apron-type thresher, using a raddle system with straw and grain rakes instead. The Eclipse could thresh a hundred bushels of grain per hour using a sweep-type horsepower. Still, the power was not dependable because the horses slowed noticeably as the load increased. The power requirements of the Eclipse thresher prompted

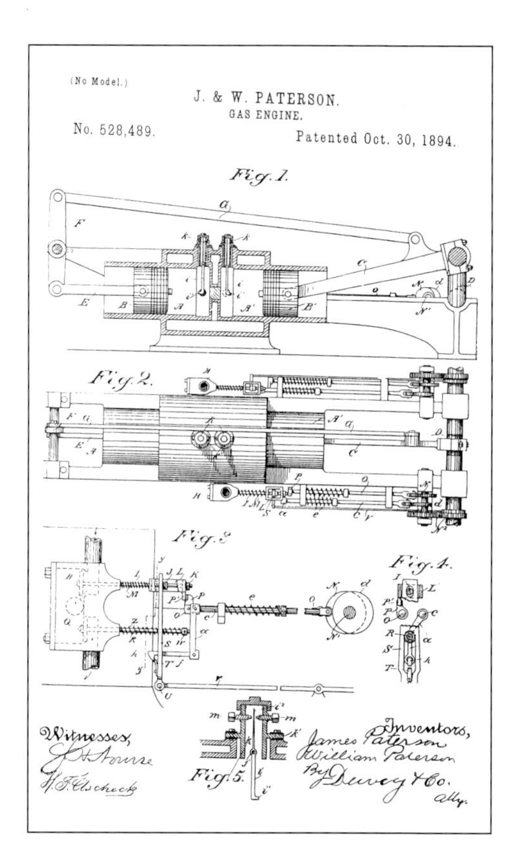

Paterson's patent drawing of 1894 shows the complex linkage that ties the two pistons to a single crank. The mechanism worked quite well, providing a balanced operation, but problems of carburetion and ignition doomed the Case-Paterson tractor project.

Case to enter the steam engine business in 1869. These engines could be used for pulling jobs, such as plowing, as well as for belt work.

By 1881, J.I. Case & Company had outgrown the partnership. A corporation was formed to replace it and named the J.I. Case Threshing Machine Company. It became known as the Case T.M. Company.

At the same time as the incorporation, Case introduced the new Agitator thresher, a machine that employed shaking straw racks. This type of mechanism was generally used in threshing machines from then on. The Agitator was an invention of W.W. Dingee, an employee and close friend of Case. Dingee obtained several other important patents for the company as well.

Jerome Increase Case died of acute diabetes in December 1892, just 11 days after he would have turned 72. Fortunately, because of the corporation form of business, the J.I. Case Threshing Machine Company was able to continue unabated. Stephen Bull, J.I. Case's brother-in-law and an original partner, assumed the presidency.

The following year, 1892 (some insist it was 1894), the J.I. Case Threshing Machine Company announced its first internal combustion tractor: the Paterson. Case had by then been in business for nearly 50 years and was the world's premier steam engine builder. By 1924, when the last Case steamer rolled out the Racine factory door, more than 36,000 had been built. Nevertheless, the company's internal combustion Paterson was not very successful.

The engine of this tractor was designed by a Californian of Scottish heritage, William Paterson, under contract to Case. William's brother, James, also was involved, but apparently did not come to Racine. The engine used two opposed pistons in one cylinder connected to one crankshaft. The rest of the machine was based largely on parts from the steam engine. The big machine produced approximately 30 horsepower for threshing and could climb an 8 percent grade. Problems with carburetion and ignition were responsible for termination of the project, however. It would be almost 20 years before Case would try it again.

After the turn of the century, several interesting events in the history of J.I. Case Threshing Machine Company took place. In 1900, branch houses were established in Europe and Russia. In 1904, Case introduced the first all-steel thresher. In 1910, the automobile line was added, as was a line of road building equipment. In 1912, the large two-cylinder gasoline tractors were introduced. Also in 1912, the Case T.M. Company entered the plow business, angering the unrelated Case Plow Works company. The last Case horsepower unit was built in 1914.

These men comprised the heart of early Case leadership and were known as the "Big Four." Clockwise from top, they are J.I. Case, Stephen Bull, Robert Baker, and Massena Erskine.

J.I. Case Plow Works

In the last half of the nineteenth century, economic cycles made it difficult for entrepreneurs to obtain a reliable flow of working capital. Men like J.I. Case were often approached to finance budding companies and technologies. Ebenezer Whiting did just that when he sought to market his "center-draft" plow in 1876. J.I. Case provided the necessary financing for the new firm, called Case, Whiting & Company. A factory was set up next door to the T.M. Company.

It was not long before financial difficulties caused J.I. Case to buy out Whiting and put his own management in place. The name of the firm was changed to the J.I. Case Plow Works. The new company was still separate from the J.I. Case Threshing Machine Company (or, since incorporation had not yet taken place, J.I. Case & Company). Without incorporation, the enterprises were kept separate to protect the assets of Case and other family members. Incorporation of the T.M. Company occurred four years later.

The year before he died, Case installed his son, Jackson I. Case, as president of the Plow Works. Upon Case's death, Jackson, hungry for cash, borrowed against his and his sisters' stock in the Threshing Machine Company. This left Stephen Bull in control. J.I. Case's son-in-law, H.M. Wallis, showed real business sense and acumen. While the Bulls looked out for the older stockholders and the J.I. Case Threshing Machine Company, Wallis looked out for the J.I. Case Plow Works and the younger stockholders, especially Jackson and his sons Roy and J.I. Case II. By 1892, Wallis took over as president of the Plow Works.

The two Case firms operated side by side without conflict for the next several decades, although the Plow Works never was as prosperous as the threshing machine company. The activities of the T.M. Company with steam engines gradually drew it into the plow business, however, and a bitter legal battle ensued. In 1912, it was decided to change the name of the threshing machine company to J.I. Case Company. Management of the Plow Works got wind of the impending change and immediately filed papers setting up a corporation under that name. In addition, the new corporation demanded that the Post Office deliver all mail addressed simply to J.I. Case to them. Protests by the J.I. Case Threshing Machine Company, who owed the bulk of the mail thus addressed, got the ruling changed.

In 1915, the J.I. Case Plow Works filed an unfair competition claim against the J.I. Case Threshing Machine Company. The T.M. Company was selling plows with the name Case on them. The suit gained much public attention as it made its way to the Wisconsin Supreme Court. It seems that the T.M. Company had acquired a great stock of plow parts in an aborted venture with a California man named Price in order to develop a large steam plowing outfit. Some of these parts were assembled and sold along with steam engines. By 1912, the T.M. Company had an arrangement with Sattley of Springfield, Illinois, wherein Sattley made plows for the T.M. Company which were sold under the Case name. The Wisconsin Supreme Court ruled that the J.I. Case Plow Works Company had exclusive rights to sell plows labeled "Case," while the J.I. Case Threshing Machine Company could use the name "Case" on products other than plows.

Some time earlier, Wallis, president of the Plow Works, had founded the Wallis Tractor Company in Cleveland, Ohio. A huge gasoline-powered monster, the Wallis Bear, appeared on the market in 1902. Wallis continued developing tractors of up to 80 horsepower, with weights approaching 30,000 pounds, until 1913. By then the firm was moved to Racine and shared housing with the Plow Works and eventually became part of the same company.

Wallis engineers, Eason and Hendrickson, came up with a new lightweight tractor design in which the under cover of the tractor was also the frame. This design used a unit or integral frame. The new tractor was called the Wallis Cub. In 1919, the Wallis Tractor Company was merged into the J.I. Case Plow Works Company.

With the introduction of the Fordson—the lightweight, low-cost tractor brought out by auto magnate Henry Ford in 1917—things were never the same in the tractor business. Ford soon garnered 70 percent of the world tractor market, driving many substantial producers out of the business, including Samson, a tractor manufacturer owned by General Motors. The little Wallis Cub held its own, for the most part, because of high quality and good design. The difficult times of the late 1920s took their toll, however, on the J.I. Case Plow Works Company. In 1928, Massey-Harris of Canada purchased the J.I. Case Plow Works Company for $4.4 million. Immediately, Massey-Harris sold the rights to the Case name to the T.M. Company for a substantial $700,000, thus ending decades of confusion and animosity.

Likeness of the original Civil War mascot, Old Abe. The eagle would sit on his perch during parades and react splendidly to martial band music. *Case Archives*

Emerson-Brantingham Company

The Emerson-Brantingham Company of Rockford, Illinois, was once a major player in the farm equipment business, rivaling the likes of Case and John Deere. Roots of the company go all the way back to 1840 when New Yorkers Pells Manny and his son, John, began developing reapers and mowers.

By 1852, the Mannys had run afoul of C.H. McCormick, whom they had bested in several reaping contests. McCormick charged patent infringement. At about the same time, John Manny contracted a company from Rockford, Illinois, to manufacture reapers. He moved the company to Rockford, taking on two Rockford industrialist partners, Waite Talcott and Ralph Emerson. Pells Manny stayed in New York.

With the help of a young lawyer named Abraham Lincoln, Manny was successful against McCormick. The trial brought much favorable publicity and the business prospered. Unfortunately, John Manny died at age 30 soon after the litigation was resolved.

After John Manny's death, the name of the company was changed from the Manny Company to Talcott, Emerson & Company. Emerson managed the enterprise on a day-to-day basis until 1895. At this time, he hired Charles Brantingham to be General Manager. When Talcott died in 1900, the firm's name was changed to Emerson Manufacturing Company.

With Brantingham as General Manager and Emerson as President, the company undertook an ambitious expansion program. When completed, Emerson-Brantingham's Rockford manufacturing facility was the largest

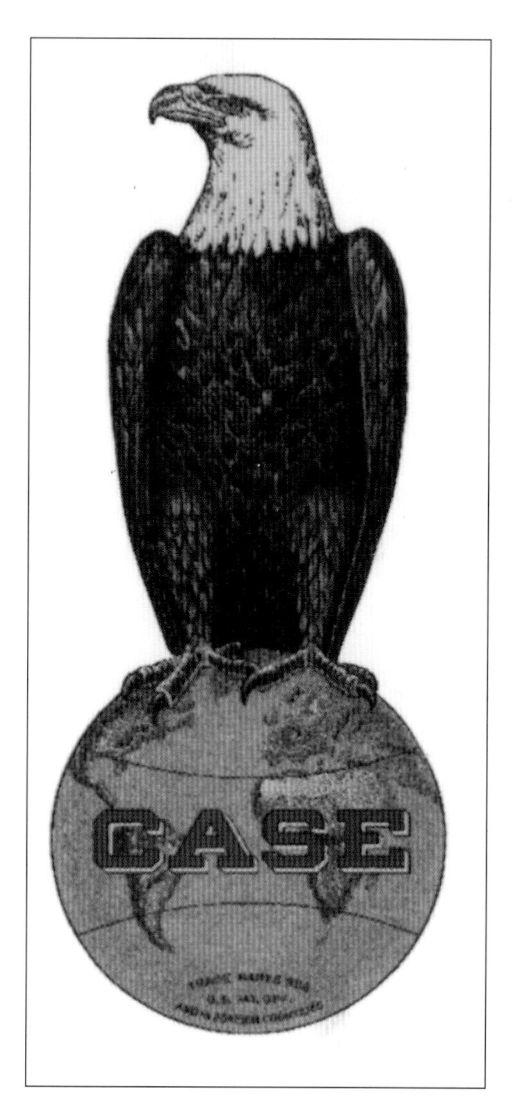

The original Case trademark showed Old Abe perched on a tree branch, but this soon gave way to the more popular design of the eagle on the globe. For a century, virtually every piece of casting produced by Case Company carried this image in relief.

in the nation, covering almost 200 acres and offering 2 million square feet of floor space.

In 1909, Charles Brantingham was elevated to the presidency, while Emerson assumed the title of Chairman of the Board. It was at this point that the name of the company was changed to Emerson-Brantingham; E-B for short. Besides their line of reaper-binders (harvesters), E-B had acquired lines including wagons and buggies, plows and tillage equipment, haying equipment, steam engines, threshers, road building equipment, sawmills, drills and seeders, and had lured the Big 4 (Gas Traction Company) gasoline tractor away from suitor John Deere.

Even though Emerson-Brantingham had acquired several steam and gasoline traction engines, the company began producing proprietary gas tractors as early as 1914. Their timing was not just right, however, as the market was moving away from both the steam tractors and the big gasoline traction engines. The Big 4 was a 21,000-pound outfit with 60 belt horsepower and 8-foot drive wheels. E-B also offered the 40-65 acquired from Reeves, a 25,000-pound leviathan. Their first proprietary tractor was the smaller Model D, loosely based upon the Big 4. The D weighed in at about 10,000 pounds and was comparable to the Waterloo Boy.

The trend was clearly toward the smaller and lighter and cheaper tractor, so E-B introduced the three-wheel Model L in 1916. This was a 12-20 machine that weighed only 5,500 pounds. There followed an improved 12-20 four-wheel tractor in 1917 and a 9-16 four-wheel tractor also in 1917. These were marginal tractors at best, due to short-lived engines and open-drive gearing. But in 1917 the famous Fordson, manufactured by Henry Ford & Son, came on the scene. It drastically altered almost everyone's concept of farm power. The capacity of Ford's marketing and the low costs of his mass production, left little room for marginal competition.

Emerson-Brantingham's product line was in much demand in the 1920s, except for the tractor line. The company employed more than 1,500 workers, who worked long shifts to meet the need for products other than tractors. Much effort was spent in design and redesign of the various tractors, but they all were uncompetitive with the offerings from Deere, International Harvester, Ford, Hart-Parr, and Case. In short, the tractor business sapped the strength of the company. The E-B company ceased to be profitable in 1920. Tractor production stopped in 1926. In 1928, J.I. Case Company bought the farm equipment division of Emerson-Brantingham, including the giant manufacturing facility. E-B's other divisions continued in Rockford, for a time.

The Case Rockford, Illinois, plant was originally built by the Emerson-Brantingham Company. It covered 200 acres and contained about 2 million square feet of floor space. When Case closed this plant in 1970, the facility was donated to the City of Rockford. Its size is impressive, even today.

Rock Island Plow Company

As far as tractor heritage is concerned, the Rock Island Plow Company has a main branch of its own: the Heider Manufacturing Company with its line of reputable tractors. Heider was formed in 1903, however, while the beginnings of the Rock Island Plow Company go back to 1855.

The business was formed in Rock Island, Illinois, by a former associate of John Deere in the plow business, Robert N. Tate. Tate's factory was only a few miles down the Mississippi River from Deere's plow headquarters of Moline, Illinois. Tate manufactured plows, choppers, rakes, planters, and hand tools for farmers of the region.

Tate was originally in the business by himself, but took in Charles Buford as

J.I. Case as a young man of 23 years. The dotted line on the map in the background traces his boat journey from New York to Chicago in 1843.

a partner after a disastrous fire in 1856. The business was then called Buford & Tate. During the Civil War years, Tate retired from the business, which was then named Buford & Company. The company prospered, but in 1882 there was another devastating fire. Buford reorganized the company with a variety of investors as the Rock Island Plow Company.

The company continued to grow and prosper into a full-line agricultural equipment company, except that neither gas tractors nor steam traction engines were offered. This changed in 1913 when an arrangement with the Heiders was made.

According to *C.H. Wendel*, brothers John and Henry Heider started the Heider Manufacturing Company in Austin, Minnesota. In 1904, the brothers moved their operation to Carroll, Iowa. Originally, a line of wooden products was offered, including eveners, whiffle trees, yokes, and ladders.

Henry Heider was mechanically inclined, however, and interested in gasoline tractors. In 1913, he obtained his first tractor patent, which featured a tractor with the engine mounted over the drive axle. A year later, in 1914, the Rock Island Plow Company made arrangements to market this tractor. Business was good—too good, in fact. Heider could not keep up with the demand and, in 1916, Rock Island bought out the Heider tractor line. The Heider name was maintained, and various tractors continued in the line. By the late 1930s, the Weyerhauser Company had taken over the company. The ravages of the Great Depression had taken its toll, and the Rock Island Plow Company was in financial difficulty. It was in 1937 that the J.I. Case Company moved in and acquired Rock Island Plow.

Chapter 2

Early Tractors

The American Civil War spawned economic transformations that had a profound impact on society during the last third of the nineteenth century. To a large extent, these transformations were disruptive to society because things were in a constant state of change. There was a marked extension and expansion of the Industrial Revolution that began with the century. New machines, power, materials, and processes continuously made the old ways obsolete. Transportation was the key. There were 35,000 miles of railroad in 1865; five times that in 1900. Pullman sleeping cars eased the discomfort of long trips. Westinghouse invented the air brake, which greatly improved railroad safety. Brayton cycle refrigeration allowed the transportation of meat and fruit by rail. Magnificent steam locomotives provided smooth, efficient rail power. The "Golden Spike" was driven in 1869, completing the first transcontinental rail line.

At the end of the Civil War, the Great Plains were largely untouched by

LEFT
This photo, dated April 4, 1907, shows in fine detail the public interest generated by steam traction machines. *Case Archives*

ABOVE
This threshing scene was very typical of grain threshing on the American farm for 100 years. Threshing continued in many smaller areas until after World War II. *Case Archives*

An early example of Case's Model 20-40. The short coupling between carburetor and intake manifold and the square radiator design indicate this unit is probably a 1913 model. *Robert Pripps*

the plow. The railroads were granted vast amounts of Great Plains land as an impetus for development of the rail lines. These railroad companies advertised for immigrants from the eastern states and Europe and offered them low-cost transportation and low-interest land loans. Thousands of railroad workers also stayed on to farm the level, fertile, and virtually treeless land. The population of Kansas, Nebraska, and the Dakotas increased sixfold between 1870 and 1890.

The center of wheat production moved from Illinois and Iowa, north and west (and these states became the big corn producing states). Gigantic wheat ranches were established in the Great Plains states, Canadian Provinces, and California. Between 1866 and 1891, the wheat crop increased from 152 million bushels to 612 million bushels.

Until well into the twentieth century, people lived in a horse-centered society. The horse served as transportation and farm power plant. The well-off farmer had a team of Clydesdales or Percherons for each 25 acres for heavy work; a fast trotter or two for his buggy; and several saddle horses. Raising and training horses were major parts of farm life. The hours of toiling together and sharing in the harvest resulted in a bonding between human and animal that transcended that which a person might have for a pet.

There was one big advantage to horse power over mechanical power. With the horse, there were no new models and no new ways to learn. Growing up on a farm, a child learned to handle the draft animals without any special training. The number of horses and mules on farms rose from 7.6 million in 1867 to 25 million at their peak in 1920.

Economic, agricultural, and industrial developments were not orderly during the turn of the century. There was overproduction, cutthroat competition, and financial panics in 1873 and 1893. Prices followed a general downward trend, forcing labor-saving methods. Hard times followed for the workers, and relief was sought in labor strikes of unparalleled violence and in political populism. Big companies got bigger. Medium and small companies often thought they were doing all right, but suddenly found they were bankrupt. Mergers and trusts were the order of the day, until the government intervened.

Steam

In this world dominated by horses, the appearance of the steam engine caused a rash of highway accidents. Even experienced horsemen found their teams difficult to control with the approach of a steam traction engine. Laws were quickly passed regulating the use of public roads, seriously restricting the operation of a self-propelled conveyance. In some cases, the engineer was required to hitch horses to the front of the machine to allay the fears of approaching horses.

It may come as a surprise to some that the very first steam power plant was invented by a scientist named Hero, who lived in Egypt in 120 BC. Hero's engine consisted of a hollow globe mounted on a pipe running to a steam kettle. Two L-shaped pipes were fastened to opposite sides of the globe. With a fire burning under the kettle, steam rushed out of the L-shaped pipes, causing the globe to rotate in the opposite direction by reaction forces.

More than 1,600 years passed before steam power would be harnessed. James Watt invented the features that made the steam engine a practical power plant. Around the turn of the eighteenth century, steam power was applied to transportation. A Frenchman named Cugnot is said to have made the first self-propelled vehicle in 1769, but it was not until 1800 that steam land machines operated in America. Robert Fulton made the first successful steamboat, the *Clermont*, in 1808.

Farm steam power, fired by the inventions of the reaper and thresher, came on the scene in 1841. Ransomes of England built the first engine. Next came "The Forty-Niner," built in 1849 by Archambault of Philadelphia. Garr-Scott followed these in 1852, then Robinson in 1860, Rumely in 1862, and Case in 1869. The early farm engines were steam power units only, with no self-moving capabilities. The self-propelling feature and the ability to pull loads began to appear in the late 1850s. By 1880, the steam traction engine was

Rear view of an early Model 20-40 tractor. Built with many steam traction running gear parts, these early tractors were quite primitive, with gears open to the dust and dirt, and an engine design that created interesting vibration problems. *Case Archives*

a mature device that found routine use in plowing, land clearing, and belt-powering a variety of farm implements. By the start of the twentieth century, 31 steam engine manufacturers produced about 5,000 engines per year.

The first Case steam engine, a portable engine of 8 horsepower, was built in 1869. The boiler was patterned after the practices initiated by railroad locomotive makers. The configuration of Case portable engines did not change much for the next 20 years. Production of portable engines continued through 1914, however, with some sophisticated designs of up to 65 horsepower being offered.

The first self-steering Case steam traction engine appeared in 1884. Although portables had been around for several decades, the Case was self-steering as well as self-propelled. It employed a drive mechanism patented by Charles Cooper of Mt. Vernon, Ohio. Subsequent traction engines were, in some cases, horse-steered and some self-steered.

Frank K. Bull, the son of Stephen Bull, became president of the J.I. Case Threshing Machine Company in 1901. From then on, Case steam tractors vastly outsold all others in a field where there were some very creditable competitors. Production continued through 1925. The largest was a 150-horsepower road engine weighing over 40,000 pounds and consuming a ton of coal per day. The largest farm engine was a 110-horsepower monster with 7-foot-diameter drive wheels.

Internal Combustion

In 1890, the Otto cycle (four-cycle) engine began to supplant the steam engine in providing nonanimal farm power. In 1902, Hart and Parr made the first production "gas" tractor and became fathers of a great industry. The automobile pioneers Duryea, Benz, Olds, and Daimler were joined by the upstart Henry Ford. By 1900, there were 8,000 auto buggies on American streets.

The first successful Otto-Cycle powered traction engine was built by John Froelich of Froelich, Iowa, in 1892. One might say this engine was a hybrid, since Froelich used a Robinson steam engine frame and running gear upon which he mounted a Van Duzen one-cylinder engine. The machine had an operator's platform in front, a steering wheel, and could propel itself backward and forward. The 20-horsepower engine operated on gasoline that, until a short time before, had been considered a hazardous by-product of the lubricating oil business. During the 1892 harvest season, Froelich employed the machine in a custom threshing operation lasting 50 days. He both pulled and powered a Case 40x58 thresher, threshing some 72,000 bushels of small grain. A year later, Froelich was instrumental in forming the Waterloo Gasoline Traction Engine Company of Waterloo, Iowa. This company went on to produce the Waterloo Boy tractor, the forerunner of the John Deere tractor line.

The early efforts of the Waterloo Gasoline Traction Engine Company did not, however, pay off in a commercially viable tractor. That honor goes to two men named Charles, of Charles City, Iowa: Charles Hart and Charles Parr. While still students at the University of Wisconsin, they formed the Hart-Parr Gasoline Engine Company, and began manufacturing engines for sale. Upon graduation, they moved their operations to Charles City and began work on the traction engine. In 1902, they completed their first unit, which was designed from the outset for drawbar work. Consequently, the transmission and drivetrain members were extremely rugged. By 1907, one third of all tractors (about 600) in America were Hart-Parrs. In fact, a Hart-Parr employee is credited with coining the word "tractor," because "traction engine" seemed too cumber-

A good example of the later design changes to the Model 20-40. Note dropped intake manifold with carburetor hanging below the frame, tin sides on radiator assembly, and full canopy on this model. The 20-40 was the subject of Nebraska test Number 7, completed May 7, 1920. It weighed 13,780 pounds for the test. *Robert N. Pripps*

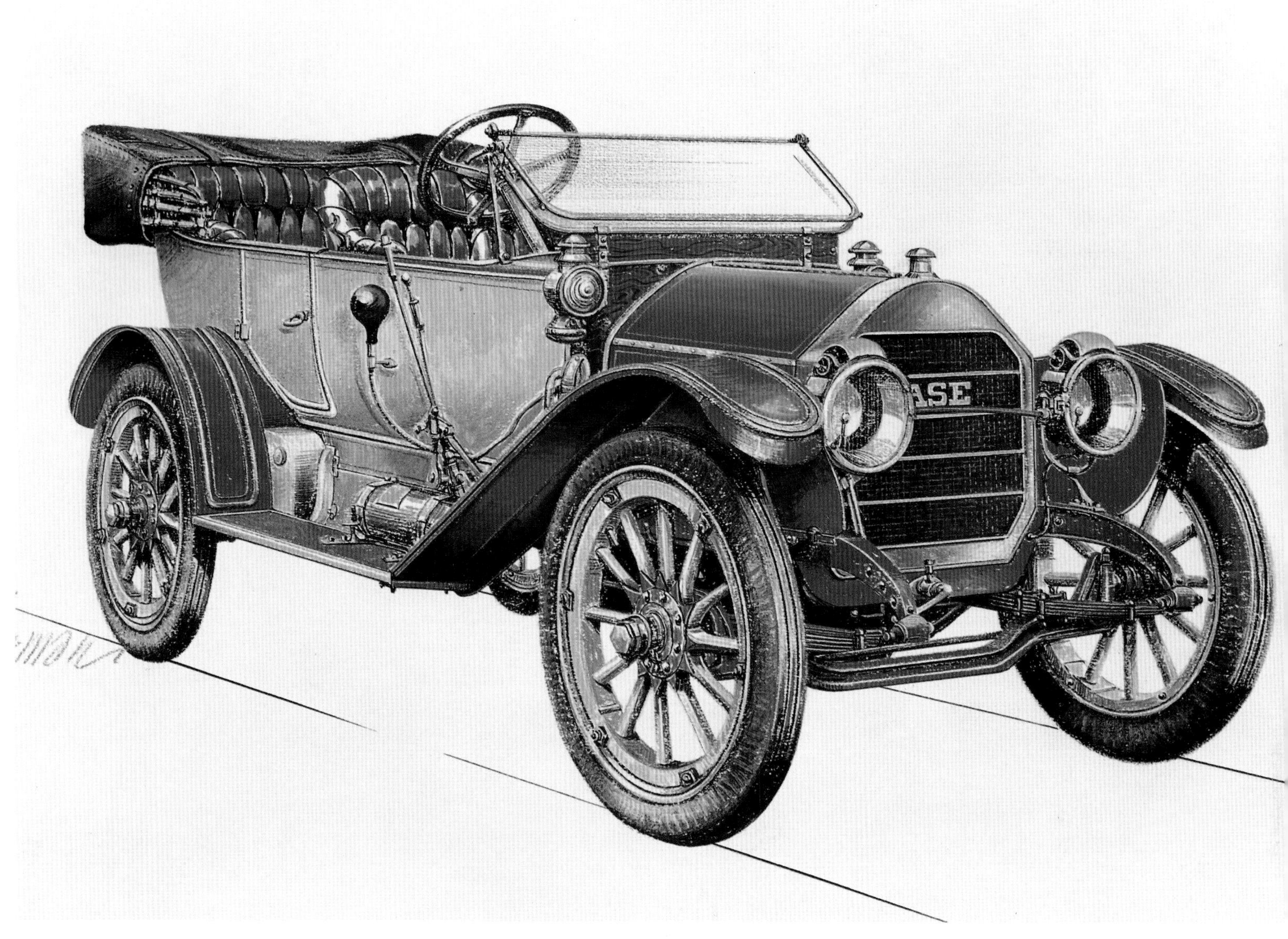

Case built a line of superb but expensive automobiles from 1912 until 1926. Shown is the 1912 Model Thirty. *Case Archives*

some for use in advertising.

The William Deering Company made its first gasoline engine in 1891. It was a 6-horsepower two-cylinder device. They also made 12- and 16-horsepower versions, and these were used on self-propelled vehicles, corn pickers, and mowers.

McCormick's first venture into the engine business was in 1897. A two-cylinder engine was made and installed on a running gear. A two-speed transmission, with reverse, was employed.

After the formation of International Harvester in 1902, interest in and work on tractors accelerated. Harvester was among the first of the long-line implement companies to offer a tractor, their first being introduced in 1906. This first unit typifies the problems of the time: its single-cylinder engine was mounted on rollers so that it could be moved back and forth to engage a friction drive. The engine had an open crank case and employed spray-tank cooling. By 1910, International Harvester overtook Hart-Parr and became the nation's number one tractor producer. Production by Harvester and the next several competitors only amounted to a few thousand tractors per year. They were expensive, being essentially shop-built, and large. Only a small market existed at the time, because most farmers could neither use nor afford such monstrosities.

Nevertheless, other farm equipment suppliers jumped on the slow-moving tractor bandwagon, and soon overproduction swamped the market.

The 150 hp "road locomotive," the largest ever built by Case.

Only a few of this very large Case engine (150 horsepower) were ever produced. Most of these were portable or skid engines. The few traction engines made developed so much torque that gear train problems were endemic. None of these engines exist today, but one boiler remains in the collection of George Hedtke of Davis Junction, Illinois. *Case Archives*

Names like Rumely, Avery, Aultman-Taylor, and Minneapolis graced internal combustion tractors. Smaller companies with good ideas also entered the fray. In 1910, Harvester added its Mogul to the lineup. Most of these tractors reflected the heritage of the steam engine; some even looked like steam tractors. They averaged over 500 pounds of weight per engine horsepower.

The giant tractors built before 1915, both steam and gasoline, were built for two purposes: driving the ever-larger threshing machines and busting the virgin prairie sod in the Canadian and U.S. Great Plains. Farm publications and Department of Agriculture studies touted the advantages of power farming. All recognized the problem: The tractor would only be viable for the average farmer when it could replace horses.

Case's 110-horsepower traction engine is possibly the most collectable engine in the world today. The size of these engines still inspires awe in spectators wherever they are demonstrated. These engines are so large that the cab and stack must be disassembled for transporting.

Case built this experimental motor cultivator during the early 1920s. None were produced for resale. This experimentation on the part of a company that was still producing large steam engines represented some creative thinking for the times. *Case Archives*

Records indicate that most families in those days actually subsisted on farms of less than 40 acres. The farther west one went, the larger the farms were, on the average.

Little Bull

To appeal to the smaller farmer, tractor designers after 1912 sought to make smaller, lighter, and cheaper tractors—machines versatile enough to replace the horse in at least some farm tasks, reducing the total number required on the farm. In 1913, an outfit called the Bull Tractor Company introduced a 12-horsepower, single-wheel-drive tractor selling for about $400. This trim, agile little device rattled the industry as it out-maneuvered the behemoths it competed against. While it was never a mechanically sound machine, it swept the field of customers, being first in sales (displacing International Harvester) by 1914. Although its popularity did not last long, it spawned a subsidiary much in evidence today: Toro, the lawn, garden, and golf course equipment maker.

As smaller, lighter tractors made the scene, an additional benefit was realized by the farmer: He could convert his horse-drawn implements to tractor use thereby saving considerably on cost. Many of these implements did not fully utilize the tractor's capability, and others were simply not strong enough to stand up, but the transition to power farming was being made.

After the abortive Paterson gas tractor experiments by the J.I. Case Threshing Machine Company in 1892, Case left the field to others until 1910. Solutions to the ignition problems were sought as time went along, but by the time they were found, the entire Paterson tractor was obsolete.

Two-Cylinder Tractors

A new machine was shown to the Case Board in 1910, and they authorized the Gasoline Traction Department and its head, D.P. Davis, to go ahead. The first to be produced was the Case 30-60 (30 drawbar and 60 belt horsepower) of late 1911. Next came the 20-40, followed shortly by the 12-25.

The 30-60 was sold through 1916 for $2,500, a considerable sum in those days, but not unlike the price of an equivalent steam tractor. It had a two-cylinder engine with side-by-side horizontal cylinders and the crankshaft running crosswise to the line of travel. It was a four-cycle engine with the pistons operating in unison, thus giving even firing once per revolution. This was the same as one of the International Harvester tractors, which was first in sales at the time, but unlike the Waterloo Boy and later John Deere tractors. The Waterloo Boy's pistons operated in opposite directions, which made the engine easier to balance but left the now-familiar "Poppin' Johnny" uneven firing exhaust note.

The Case 20-40 was offered in 1912 and built through 1920. In 1913, it won two gold medals in the Winnipeg trials for fuel economy. The engine of the 20-40 was a horizontally opposed two-cylinder unit, unlike the Paterson machine which was of the opposed piston type.

The 12-25 was introduced in 1913 in response to the lighter weight Bull tractor and the Wallis Cub. At 9,000 pounds, the 12-25 was considerably heavier than these. Its price of $1,425 was higher as well. The 12-25 used a two-cylinder horizontally opposed engine with a cooling pump, fan, and radiator. It also had an enclosed two-speed transmission and bull-gear final drives. Later tractors had a cylindrical kerosene tank over the right rear wheel and a similar water tank over the left. The 12-25 was produced through 1918.

In 1916, Case stockholders created the position of Chairman of the Board to which Frank Bull, the son of Stephen Bull, ascended. Warren J. Davis, who had been treasurer, assumed the presidency.

CASE

Chapter 3

The Crossmotor Era, 1916–28

During World War I, Case did not build equipment or ammunition for U.S. forces. They did, however, make under license a 130-horsepower Clerget nine-cylinder rotary aircraft engine for the British Sopwith Camel fighter. This was not Case's only venture into the aircraft business. They made their own airplane prototype prior to the war, but nothing ever came of it.

World War I ended on November 11, 1918. Tractor production had been brisk. From a total of 25,000 tractors on U.S. farms at the end of 1915, there were more than 85,000 at the end of the war. Production continued unabated through 1920, with Henry's Fordson garnering 70 percent of the market. By 1921, an economic slowdown precipitated by the end of war production led to real problems for the farm equipment business. To make matters worse, it seemed as if the entire American farm community had decided at once that the big, expensive tractors—especially the steam tractors—were things of the past.

LEFT
Case's Model 15-27 was built from 1919 through 1924. A medium-size tractor that could power a thresher or pull a plow, it became the largest seller of the Case tractor line during its production run.

The Wallis Cub Junior first appeared in 1915 and was marketed by the J.I. Case Plow Works, a completely different firm than the J.I. Case Threshing Machine Company. In 1928, the Plow Works was bought by Massey-Harris Company of Canada, and the Case name was sold back to the T.M. Company.

In 1921, Henry Ford saw his sales beginning to slip and lowered his price from $795 to $625 to $395 throughout the year. This caused other manufacturers to enter a price war that decimated the rolls of the tractor makers.

Meanwhile, back in Racine, the J.I. Case Plow Works was producing their creditable Wallis Cub Junior, introduced in 1917. It was a redesign of the Cub of 1914; the redesign eliminating two of the Cub's shortcomings: too much weight and exposed final drive gears. The Cub J, or Junior, weighed in at only 4,000 pounds. The J experienced some sales resistance because of its three-wheel design, so in 1919 a conventional four-wheeler, the Model K, replaced it. This was followed by the Model OK in 1922 and the 20-30 in 1927. The economic hard times of the 1920s, however, brought the Plow Works to its knees. In 1928, Massey-Harris took over.

The First of the Crossmotors

The next tractor from Case was the four-cylinder Model 10-20. Its engine was taken from the Case automobile, although the engine was redesigned after production began. That redesign was an important one, as it produced an engine that would prove to

Beginning in 1919, Case built the Model 22-40, a large tractor that would perform with the larger steam traction engines. Weighing almost 5 tons, these tractors could pull five bottoms in average soil. Owner: John Davis of Maplewood, Ohio.

be the powerplant for the future for Case, the Crossmotor.

The Case car business is a curious story. By 1910, the Pierce Motor Company (not the same one that made the famous Pierce Arrow) of Racine was principally owned and managed by senior Case people. When financial difficulties were encountered in 1911, Case agreed to market the entire production of Pierce autos through their dealer organization. The intent was to beat the competition to the vast rural market. By 1912, Case bought the Pierce Company and put its name on the car. Thus, Case was in the automobile market for the next 15 years, making what were generally considered to be good quality, but expensive, cars.

The redesigned engine for the Model 10-20 tractor that replaced the car engine used for early production was a vertical four-cylinder overhead valve type known as the Crossmotor. The upper half of the crankcase and the cylinders were cast enbloc. The cylinder head was removable, and there were large ports in the crankcase to allow cleaning of the passages. This type of engine would characterize Case tractors for years to come.

This advertisement provides a glimpse of the bitter battle that went on between the two Case companies. Note the text on the bottom of the ad that reads, "Our plows and harrows are NOT the Case plows and harrows made by the J.I. Case Plow Works Company."

The fuel tank on top of the left fender held gasoline, which was used to start the engine. Once the engine was warmed up, it was switched over to either low-cost distillate fuel or kerosene.

The Model 10-20 tractor was unique in another way from its previous stablemates: it was a three-wheel machine and used an unusual drive. In normal operation, only one rear wheel was driven by the engine. Under a heavy load, the other rear wheel could be clutched in. The single-wheel drive eliminated the need for a differential. The 10-20 sold for about $900 and weighed about 5,000 pounds.

Model 9-18

Fortunes at the J.I. Case Threshing Machine Company were somewhat better than those of the Plow Works. D.P. Davies was in charge of development from 1910 on. His genius for things mechanical gave him a more or less free hand with the Case Board.

Following the successful work done with the lightweight Model 10-20, Davies was given the go-ahead for a new departure in tractor design: the Crossmotor. The first of these, new for 1916, was the diminutive Model 9-18 that weighed just 3,800 pounds.

The 9-18 was a conventional four-wheeled tractor built in the war years of 1916, 1917, and 1918. Almost 5,000 were sold. Its 236-cubic-inch four-cylinder engine was governed to run at 900 rpm.

In 1918, with the onslaught of the Fordson looming, Case replaced the 9-18 with the 9-18B. It was essentially the same tractor, but with a one-piece cast iron unit frame like the Fordson rather than the steel fabrication used previously. The frame included the engine block. It mounted the front wheels, and included the back axle and everything in between.

Still in 1918, Case engineers upped the engine speed 150 rpm and re-rated the tractor as a 10-18. Interesting features of the 10-18 engine were a sight gauge for oil circulation and a float-type oil level indicator. The 10-18 also featured a water pump and thermostat arrangement known as the Sylphon System. A Case-patented air washer intake filtration system was also carried over from the later 9-18s. The 10-18 was sold through 1921.

Nebraska Tractor Test Law

In April of 1920, the 10-18 became the first Case tractor tested by the University of Nebraska. Problems with the quality of early tractors and outrageous advertising claims prompted the establishment of a testing facility at the University, and the University of Nebraska test became the benchmark by which tractors were rated. The Case Model 10-18 was the third tractor to be tested by the University of Nebraska.

About 200,000 tractors were sold in 1920, but the war boom was over and sales began to drop off. Manufacturers, however, did not know to adjust their output. Domestic sales in 1921 amounted to only 35,000 tractors (half of which were Fordsons), and the indus-

To the left and above the belt pulley in this photo is the lubricator, a unit filled regularly with oil that fed individual engine bearings in a drip system. A small thin belt is connected from the engine to this lubricator. The system used the oil once and discarded it into the sump.

The powerplant of the Model 22-40 displaced 641 cubic inches. Cylinders were cast in pairs and could be removed from the block for servicing. The crankshaft was 3.5 inches in diameter, and the engine ran at a governed speed of 850 rpm.

A 1921 *Motor Age* advertisement shows a 9-18B driving a thresher. Notice the thresher is being fed from both sides, which would make quite a load for the tractor. Notice also that drive belts were not crossed for Crossmotors driving Case threshers.

try was devastated. Some of the drop in sales was due to falling crop prices, but much was due to the fact that farmers were becoming increasingly wary of unscrupulous tractor makers.

Defective tractors, overly enthusiastic advertising, and "paper" tractor companies had caused the farmers to clamor for consumer protection even before the boom war years. In 1915, *Power Farming* magazine had called for the standardization of horsepower and capacity ratings, but the industry did not respond. The magazine continued to editorialize for a national rating system, saying that the states would act if the federal government failed. State tests, they said, would result in a hodgepodge of requirements.

The federal government did not react, so in early 1919 a Nebraska legislator, Wilmot F. Crozier, introduced a bill that would require all tractors sold in Nebraska to be submitted for testing. Crozier was also a farmer who had already had poor experience with oversold and under-designed tractors. Crozier's bill provided for the testing to be done by the University of Nebraska. Another aspect of the bill was that manufacturers would be required to keep an adequate stock of spare parts in the state.

The Nebraska Tractor Test Law went into effect on July 15, 1919. The Nebraska program was so comprehensive that other states abandoned their own plans and relied on Nebraska's results, as did much of the world.

Testing began in the fall of 1919, but an early snowfall interrupted the activities until the spring of 1920. The first tractor to successfully qualify for sale in

The Model 12-20 Crossmotor replaced the earlier 10-18 and was slightly larger. Most characteristic of this model are the stamped steel wheels that replaced the traditional spoke design. Owner: John Davis, Maplewood, Ohio

This Case Model 25-45 is an up-rated version of the 22-40. The 25-45 was a big machine, weighing in at more than 10,000 pounds. Note the eagle trademark between the CASE letters on the radiator. This usually designated the up-rated versions of Case's Crossmotor series as opposed to the plain CASE name in larger letters in the casting. *Robert N. Pripps*

Note fuel tank in front of engine, next to radiator, and starting fuel tank at hood side. This model was quite compact and rugged. Also new for the 12-20 model was a color change from the traditional green with red wheels to all gray. This change would affect Case's entire tractor line.

The Model 10-18 supplanted (and soon replaced) the 9-18. Few differences existed between the two. Most notable of these was 100 rpm higher governed speed, which added more torque and power. Owner: John Davis, Maplewood, Ohio.

Because of their more interesting color, the earlier series Case Crossmotor tractors are more collectable than the later all-gray tractors. Pin-striping was still a cultivated art in the late 1910s, and many examples of this are evident on these models.

Case's Model 9-18 tractor was the first four-wheeled Crossmotor design. The unit used an engine with only two main bearings, an engine also used in Case combines of the times. Owner: Bruce Tadlock of Ironwood, Michigan.

Nebraska under the new law was the Waterloo Boy Model N. Since that time, the Nebraska Tractor Tests have been the common yardstick by which tractor performance is measured.

Model 15-27

November 1918 saw the third of Case's famous "Crossmotors," the Model 15-27. Production ran from 1919 to 1924. The 15-27 used the same type of cast unit frame as the 10-18. A 382-cubic-inch four-cylinder engine, turning a nominal 900 rpm, powered the machine. The air washer was standard, as was pressure lubrication of critical parts. Everything about the 15-27 was robust. The three-bearing crankshaft was 2.5 inches in diameter; the rear axles were 2.75 inches. Interestingly, some 15-27s were provided with two pivot holes for the front axle. When the off-center hole was used, the right front wheel could run in the furrow as a steering guide. The two rear wheels ran on unplowed ground for better traction. If not equipped with this offset axle, a "plow guide" could be attached to the front wheel.

These and other features brought the weight of the 15-27 up to 6,500 pounds and the price to $1,700. The farmer's loyalty to Case quality was severely tested since the $395 Fordson, with a high-compression head and gasoline fuel, could produce the same power (although maximum drawbar pull would be almost a thousand pounds less for the Fordson). Nevertheless, more than 17,000 Model 15-27s were sold over its five-year production life. Only 17 were sold in 1921 however, the year of the tractor price war.

In 1925, the 15-27 was re-rated and re-identified as the 18-32. About the only change was increasing the rated engine speed to 1,000 rpm. Production continued through 1928; however, in that last year, it was called the Model K. Almost 15,000 of this series was produced. When combined with the 17,000 Model 15-27s, this series was the best-received of the Case Crossmotor line.

Model 22-40

Built to provide power for belt loads like the Case 32x54-inch thresher, the Model 22-40 was a big tractor when introduced in 1919. Ad literature of the

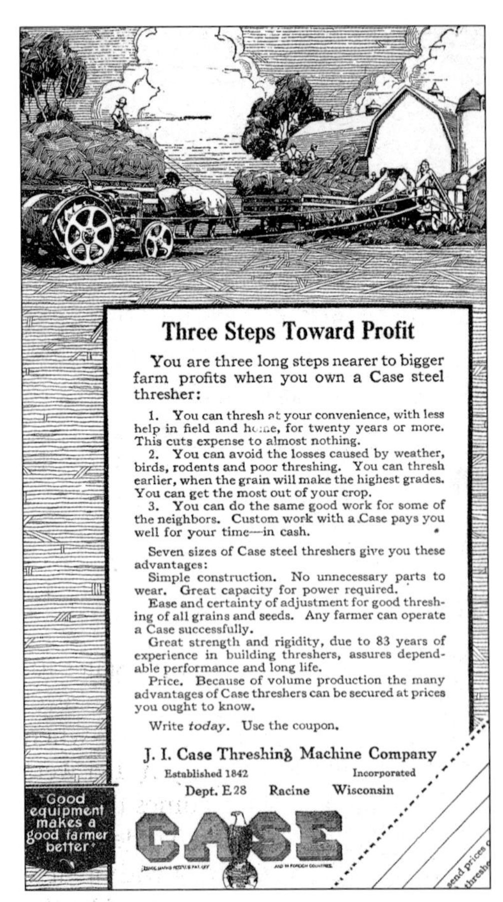

This 1925 ad by the J.I. Case Threshing Machine Company features a 12-20 tractor and a small Case thresher, possibly a 20x28.

A total of 17,628 Case 15-27 Crossmotors were built between 1919 and 1924. The 15-27 weighed in at about 6,500 pounds. This one is a 1920 model. Owner: John Davis of Maplewood, Ohio.

day said the 22-40 was for "Any farmer whose work is heavy enough to require this much power."

The 22-40's weight of almost five tons and torque of the massive 641-cubic-inch engine required a stronger frame than Case's typical cast units. The new frame was a hot-riveted steel frame with 8-inch channels. Boiler-plate was added for increased rigidity. The four cylinders were cast in pairs, rather than all four in one block, as was Case's usual practice. The crank-shaft was 3.5 inches in diameter. The engine was rated at 850 rpm.

The 22-40 was a steady seller for Case up through 1924, with a total of 1,669 being sold. Even in 1921, when the tractor market hit bottom, 181 were sold, even at almost ten times the $395 price of a Fordson. The Model 22-40 was often purchased along with a large threshing machine and was used by the custom threshers of the day. The result was a rig that cost more than $6,000, a significant amount of money back then. Just as today, the price of heavy equipment was mind-boggling, but big machines meant big production and fast payoffs.

In 1925, a variation of the 22-40 theme was introduced. It was the same as the 22-40 except for the type of carburetor and the magneto. It was identified as the Model 25-45. This model was produced through 1928, but in the final year it was called the Model T. Approximately 980 25-45s and Model Ts were sold.

Model 12-20

The 12-20, an updated version of the earlier Model 10-18, came out in 1921. Production continued through 1928, but in 1928 the name was changed to the Model A. Although the 12-20 looked much the same as the 10-18 it replaced, it was in fact longer by 7.5 inches and heavier by 700 pounds.

The Case 15-27 Crossmotor was powered by 382-cubic-inch four-cylinder engine, rated to run at 900 rpm. In 1925, the 15-27 rating and designation were changed to 18-32.

The Model 18-32 was the up-rated version of the earlier 15-27 and sported only a few changes, most of which were improvements in the horsepower department.

To handle the additional power, the internally expanding clutch was replaced with a disk clutch. Also, the engine bore was increased, and a center main bearing was added.

The most distinctive feature of the 12-20 was pressed steel wheels. These were cut from plate steel and hot riveted to the rims. The triangular cutouts for the spokes left quite a stylish wheel. The 12-20 introduced the all-gray paint scheme, differing from the previous green and red coloring. Striping remained the same as before.

A total of 11,414 Model 12-20s and Model As were delivered at an average price of $1,100 each.

Model 40-72

The big one! The Model 40-72 Crossmotor weighed in at about 22,000 pounds and stood over 9 feet tall. It was built along the lines of the 22-40, with the riveted frame and cylinders cast in pairs. The engine displaced an immense 1,232 cubic inches. The kerosene tank held 52 gallons—the 40-72 could easily consume this in a normal day's work. A normal day's work included plowing with a 12-bottom (14-inch) plow. Only 41 of these $4,100 tractors were built between 1920 and 1923; none in 1921. Seven or eight are known to exist today.

Industrial versions of the 12-20, 10-18, and 15-27 were offered. About the only difference between these and the farm versions was the use of hard

RIGHT
The large arm that protruded from this model's right side frame was the support for the clutch assembly, housed inside the belt pulley. Gearing was simple, but enclosed, a positive improvement.

CASE
CASE
RAISE TO
CREASE

The Case Crossmotor production line of 1927. In foreground is a Model 18-32. Next is a 12-20 with scalloped wheels. By this point in time, Crossmotor series all had a model designation. These two would have been Models K and A, respectively. *Case Corporation*

Leon R. Clausen was Case president from 1924 until 1948, and probably had more influence on the company during his tenure than anyone since J.I. Case.

rubber tires, different gear ratios, and lower ground clearance. Special road wheels were offered for the 15-27/18-32 and the 22-40/25-45 for highway work. These wheels, with heavy cleats, added 1,000 and 3,000 pounds of weight. Rice field and orchard variations were also offered.

Leon R. Clausen

The year 1924 was pivotal in the history of the J.I. Case Threshing Machine Company. In June 1924, Leon R. Clausen left the employ of Deere & Company and assumed the presidency of J.I. Case Threshing Machine Company. No one, save J.I. Case himself, had more influence on the fortunes of the company for both good and ill.

Clausen was born in nearby Fox Lake, Wisconsin, in 1878. He was the son of a Danish farmer, who was part owner of a Great Lakes vessel and owner of a grain elevator where Leon worked as a young man. He eventually became part owner of the elevator.

In 1897, Clausen graduated from the University of Wisconsin with a degree in electrical engineering. He held a variety of jobs in industry, eventually becoming a railroad signalman, superintendent of the signal department of a major railroad, and President of the Railroad Signal Association of America.

During Clausen's early thirties, he became interested in the Progressive movement, led by Wisconsin's Robert LaFollette, that opposed the alliance of wealth and government. Clausen was irate over the involvement of the government with the railroad magnates. His ire eventually caused him to leave the railroad business and take employment with Deere & Company in 1912. The farm equipment industry was, at that time, relatively free of government intervention.

In 1912, Deere had just consolidated related businesses under the corporate banner. Clausen was made manager of Deere's Dain Manufacturing Company of Ottumwa, Iowa, a manufacturer of haying equipment. By 1919, Clausen had become Vice President of Manufacturing Operations for Deere & Company and a member of the Deere Board of Directors. Morgan Stanley, Inc., which handed the stock for Case, approached Clausen about the job at Case. Morgan Stanley could see that new blood was needed at the top to get things going again.

To say that Clausen was conservative would be an understatement. He was a rigid authoritarian manager who was little interested in what customers thought they wanted. He, and Case, would tell them what they needed. Under Clausen, the company headquarters resembled a Dickensian counting house with tall open desks, men on stools wearing white shirts and green eye shades, and spittoons on the hardwood floors.

In 1924, Clausen knew what Case needed to get out of the doldrums of the times, however. One of his first acts was to terminate the manufacture of the automobile. The car had become an unnecessary diversion of cash and talent from the primary line. Ford, with his Model T, had clearly won the hearts and minds of the rural folks anyway.

Next, he phased out the manufacture of steam engines. The last of 35,737 engines, an 80-horsepower portable, was built on September 28, 1924. It was not shipped until 1927, and it went to South America.

Finally, and of great competitive importance, Clausen authorized engineer D.P. Davies to modernize the entire tractor line. Much of the competition had already introduced new models and lines, like the John Deere Model D which looked more like the Fordson. The new Case line would not be ready until 1929.

The Fordson Phenomenon

On June 28, 1914, in the city of Sarajevo, terrorist Gavril Princips assassinated the Crown Prince of Austria, Archduke Franz Ferdinand. For presumably harboring a terrorist organization, Austria-Hungary declared war on Serbia. Germany, Russia, France, Turkey, Bulgaria, Belgium, and Great Britain were soon dragged into the conflict, which became known as World War I. The United States entered the conflict on April 6, 1917.

The low cost of foods that had been imported to Britain from America and Russia caused British farmers to turn to livestock production. When the Turkish Navy blockaded the Dardanelles in 1914, they effectively shut off the flow of wheat from Russia. Later, the German U-boat threat began to curtail grain shipments from the other sources. The British government established tilling goals for arable acreage, but records indicate there were just 500 tractors in Great Britain in 1914. To achieve their tillage goals, the British Board of Agriculture ordered tractors from British makers to the limit of their production capacity. They also imported all the Waterloo Boys, International Harvesters, and others they could get—but still food production fell short.

Lord Percival Perry was then chief of Ford's British subsidiary and a member of the Board of Agriculture. Lord Perry arranged for testing of a prototype tractor being developed by auto magnate Henry Ford. In May 1916 a panel of five judges, all experienced agriculturists, was favorably impressed by the tractor's durability, ease in starting and handling, and small size and light weight. They recommended that the British Ford subsidiary go into immediate production. Since Ford had already been granted a license to set up a factory in Cork, Ireland, originally to produce the Model T, that sight was chosen for the tractor. Henry Ford was delighted and generously made a gift of the drawings and patent rights for the duration of the conflict, not asking for royalty payments.

Thus, Henry Ford was in the tractor business.

The Fordson tractor was the product of a company owned by Henry Ford and his son, Edsel. To avoid hassles by the auto company's stockholders, the Fords incorporated a separate company to make the tractor: Henry Ford & Son. (Transatlantic cable abbreviations led to the adoption of the Fordson name.) Ford engineer Eugene Farkas designed the Fordson using the unit frame concept pioneered by Wallis. It was made to be mass-produced (almost a million were made in the next 10 years), light (2,700 pounds), cheap (as low as $395 in 1921), and powerful (it had a 10-20 rating). In contrast, Emerson-Brantingham's 9-16 sold for $1,125 before it was discontinued in 1921. John Deere's Waterloo Boy and International Harvester's Titan were selling for around $1,200. The three-wheel Case 10-20 was priced at $900. The 10-20, the first of the Case four-cylinders, had been introduced in 1915. It weighed over 5,000 pounds and sold for $900.

In 1927, Ford ended United States production of the Fordson, saying he needed the facility for the new Model A car. Production continued in Ireland and England, but the Fordson was never again such a dominant United State's competitor.

CASE
CASE

Chapter 4

General Purpose Tractors, 1929–39

The early 1930s brought many changes. With the coming of the Great Depression in the Fall of 1929 came a severe drought in the Great Plains. The once prosperous wheat-producing lands were largely abandoned. As the Depression spread, farm income dropped worldwide and tractor production plummeted. In 1932, tractor sales in the United States dropped to about 19,000 units; the least since 1915. This decimated the tractor manufacturers. Of the 47 tractor makers left after the Great Tractor War of the 1920s, when Ford dropped prices to below cost, only seven of any substance remained producing wheel-type farm tractors in 1933. These were (ranked by size):

1. International Harvester Company
2. Deere & Company
3. J.I. Case Company
4. Massey-Harris Company
5. Oliver Farm Equipment Company
6. Minneapolis-Moline Power Implement Company
7. Allis-Chalmers Company

LEFT
The Model C was also produced in a row-crop version, designated CC. The 1938 model pictured above has factory cast wheels for rubber tires. Cast wheels of this type were deliberately made of heavy construction to add weight for improved traction. Owner: Jay Foxworthy, Washburn, Illinois.

ABOVE
The Case Model C was introduced in 1929, about the same time as the Model L. Both models were originally offered on steel wheels, for their introduction predated the agricultural use of rubber tires. The tractor pictured above was converted from steel to rubber at some point in time. Wheels of this type are referred to in the hobby as "cut-offs" and are often good examples of careful shop retrofitting.

This 1934 ad shows a Model CC. The CC used a unique steering linkage that protruded over a foot from the left side of the radiator. Dubbed the "chicken roost" by the competition, Case farmers who used this model quickly learned to make only left turns when approaching fence rows, due to the extra distance required when turning right.

In 1934, tractor sales increased 40 percent over 1933, so the survivors were optimistic. Research and development continued unabated. Since 1916, the most dramatic changes with Case tractors were three new models and the switch from green and red to the all-gray paint. The company needed some new product to survive. Case's sales manager, G.B. Gunlogson, recognized the changing market conditions. Many farmers were not buying the old-type tractors because of their limited capabilities. If they had the power to plow, they were too big for cultivation of row crops. If they were high and light for cultivating, they could not plow. Most farmers just kept their horses.

In 1921, at Gunlogson's suggestion, Case built and tested an experimental motor cultivator. Hard times of the early 1920s forced the company to abandon the project, although valuable data was gathered for the future.

When Leon Clausen arrived on the scene in 1924, he recognized that the tractors Case offered were obsolete. From his days at Deere, he knew the John Deere Model C (later to be called the GP, for General Purpose) was in the works. It would be released to the public in 1928. Yet Clausen was not ready to abandon the traditional standard-tread tractor. When he ordered modernization of the line, he insisted that two new tractors, the Model L and the C, be released at one time, replacing the entire Crossmotor series. The Model C would also be built as a general purpose machine, identified as the Model CC. Accordingly, Clausen ordered a full set of custom implements to be developed simultaneously.

Model L

The Case Model L tractor was designed to replace the 18-32 and 25-45 Crossmotors. It was the size and weight of the 18-32 but had the capabilities of the 25-45. The Model L, probably by design, was superior to the John Deere

Engine design of all-gray-era Case series tractors included a cast oil pan, large "hand holes" for gaining access to crankshaft bearings, and a wet-type clutch that ran in motor oil.

A Model CC pulling a Model A-6 combine. Note how badly lodged this oats crop has become. Note that the rear tractor tires are on backwards. *Case Archives*

Case's Model RC was the smallest of the gray-era series. Over-the-center steering was used for the first two years, a vendor-purchased unit that was given up due to cost considerations. Most over-the-center RC units had a single front wheel, which worked fine for cultivating row crops. Dual front row-crop and wide adjustable front axle configurations were offered, but both are a rare item today.

The Model RC was a light, row-crop tractor with adjustable rear tread width, standard power take-off, a three-speed transmission, and steering brakes. Rear wheels on the unit pictured above are "cut-off" steel adapted for rubber tires. Rear tires on the RC were originally 9R10x36-inch size. Tires on this unit are unusually large.

A Model L pulling a 10-foot WL series wheatland disc plow and double-fluted seed drill. Special hitches of this sort became popular in the grain belt as a time-saving device. *Case Archives*

A correct example of an original 1929 Model C. Note the absence of extruded lettering on the hood sheet metal, typical of all early gray era production. Early gray units used decals on hood sides. Power for the R series Case tractors was provided by vendor-built Waukesha L-head engines, using oil pans of "hand hole" design and wet-type clutches furnished by Case.

Both the C and L models from Case were originally produced with flat-type fan belts and Kingston carburetors. V-type belts and Zenith carburetors were introduced to both models during the early 1930s. Unit pictured above is correct in both examples. Owner: John Davis, Maplewood, Ohio.

Model D of 1927 in almost every category. The Case Model L, when tested at the University of Nebraska in 1929, weighed 5,307 pounds and produced 44.01 horsepower on the belt. Drawbar horsepower achieved was 30.02. Fuel consumption at rated drawbar power was 6.52 horsepower hours per gallon of kerosene. Deere's Model D weighed 4,917 pounds and produced 36.98 maximum belt horsepower when tested in 1927. Drawbar horsepower was 28.53. Rated drawbar fuel consumption was 6.69 horsepower hours per gallon of kerosene.

The Model L was designed to be rugged, versatile, and dependable. It was built using the unit or integral-frame concept pioneered by Wallis. It had a four-cylinder overhead valve engine that displaced 403 cubic inches and was rated at 1,100 rpm (the Deere Model D's two-cylinder engine displaced 501 cubic inches and ran at 800 rpm). The Model L used a roller chain final drive, as did the John Deere D. The Model L sported a new three-speed transmission, while the D relied on a two-speed unit until 1935. The Model L featured a rear PTO for driving the new shaft-powered binders and combines. The John Deere D did not at that time.

The Model L was built without much change through 1940. About 32,000 were built for agricultural purposes, plus about 2,100 for industrial and special applications. It was priced at $1,295 in 1930 (about the same as the Deere D). Rubber tires were offered in 1934, and with the Model LI came dual turning brakes. Like most of the tractors of the time, electric starting was not found on the Model Ls until late in the tractor's production, although industrial versions received them earlier.

Model C

Case brought out the Model C, which was a scaled-down version of the Model L, in 1929 at about the same

Using the higher ground clearance row-crop Model CC coupled to the wide standard front axle produced the Model CC4. Many CCs were sold new with both front ends, wide for plowing and row crop for cultivation.

time as the Model L. Its 324-cubic-inch four-cylinder engine, like the Model L's, had removable cylinder liners, a three-bearing crankshaft, full pressure lubrication, and an air cleaner. Also like the Model L, the Model C used chain drives to the rear wheels, and it too had a three-speed transmission. The Model L was rated as a three- to four-plow tractor, while the Model C was rated as a two- to three-plow machine.

While the Model L was designed to compete with the John Deere Model D, the Case Model C was designed with the Fordson in mind. Clausen, in a list of things he wanted to see in the small tractor, insisted that balance be such that tipping backward would not be a problem, which it certainly was with the Fordson. He also proposed a target weight of 2,800 pounds, like the Fordson, but the Model C turned out to be over 4,000 pounds.

Once banished to Ireland, the Fordson underwent some redesign before it went on sale again. The power had been increased through an eighth-inch increase in bore diameter; a high-tension impulse magneto replaced the old flywheel-coil system that caused so much hard starting. A water pump had been added, and other miscellaneous improvements were incorporated.

In 1929, the Model C was tested in Nebraska and produced 29.81 belt horsepower and 19.6 drawbar horsepower. At rated power, the Model CC produced a respectable 11.06 horsepower hours per gallon of kerosene. The Irish Fordson made 29.09hp on the belt and 20.69hp drawbar. Rated fuel consumption was 5.91 horsepower hours per gallon of gasoline.

Because of a peculiarity in the Tariff Act, Fordsons could be imported duty-free. Before the Depression, Fordsons were coming into the United States at the rate of 1,500 to 2,000 per month. The sales of the Model C were more on the order of 300 per month.

Although the stock market crashed in the fall of 1929, the dire effects of the ensuing Depression were not felt on the farms until 1931. Tractor production in 1930 was about at the average of 200,000 per year by an amazing 38 companies. The 1920s and 1930s would see great changes in what was considered a conventional tractor configuration. The greatest interest was in the general purpose, or all-purpose, machine that could replace the costly and inefficient horse. Nevertheless, in the great wheat lands of the West, the standard tread plowing tractor held its own. Gradually, the general purpose machines outsold the standard treads, and manufacturers that did not offer a general purpose machine were soon left behind. Case was not to be left behind. Its first row-crop offering, the Case Model CC was ready for testing at Nebraska in late 1929, and went on sale as a 1930 model soon thereafter.

The CC4 used the same front axle as the standard Model C, with a riser block between the axle and radiator support. Some units also used raised spindles, which placed the wheel center higher in relation to the king pin.

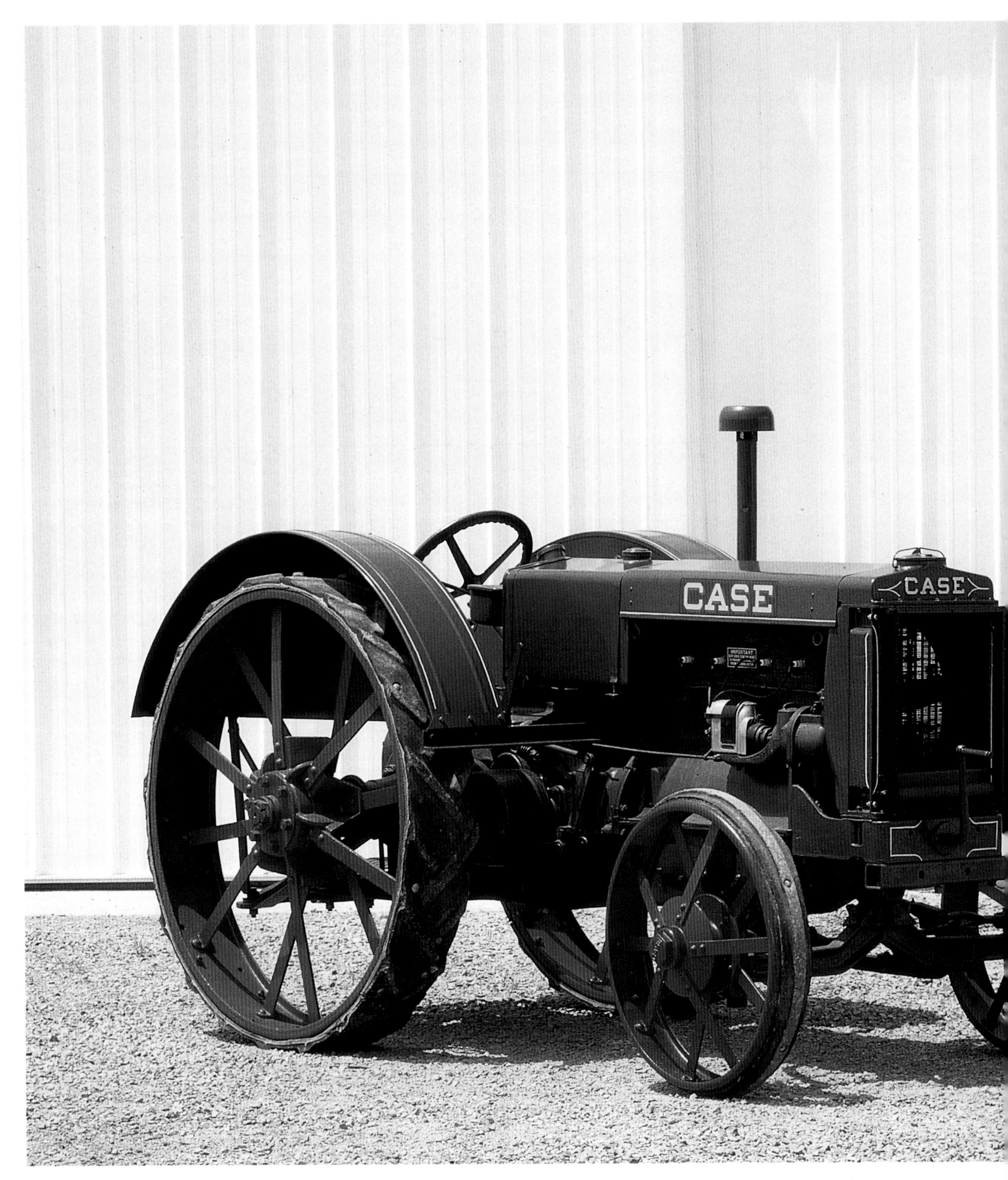
CASE
CASE

Model CC

The Model CC was a general purpose machine in every sense of the definition. It had crop clearance, a rear PTO as well as a belt pulley, adjustable rear wheel spacing, a narrow front end (although a wide front was optional), and turning brakes. It was capable of pulling two 14-inch plows and a variety of custom implements, such as mowers and cultivators. Under the skin, it was the same tractor as the Model C. Sales in 1930, however, did not match those of the Model C. Case salespeople clamored for the CC to be made lighter and cheaper. The 4,200-pound Model CC was not given a power rating by the company, but during the Nebraska tests its rated power was a little over 17 horsepower on the drawbar and 28.7 horsepower on the belt. The Model CC was more powerful than the competition, but it really was about the same weight. The Farmall, for example, weighed about 4,100 pounds and, although not rated either, was considered to be a 10-20. Clausen's reaction to the salespeople's request for change was typical of him: "A company cannot always be changing its policies in reaction to competition. If it does, it will never be a leader."

The year 1930 also saw the first GP offering from the newly formed Oliver Farm Equipment Company: the Hart-Parr Oliver Row Crop 18-27. Massey-Harris, operating from the old Plow Works facilities, was next with their unusual four-wheel-drive General Purpose 15-22. Twin City introduced their "Kombination" KT 11-20. The next several years saw a dozen more entries in the GP field, including the John Deere Models A and B, the Allis-Chalmers All-Crop, the larger Farmall F-30 and the smaller Farmall F-12, plus offerings from Huber, Minneapolis-

Because the CC4 is an adaptation of the CC, this model also used turning brakes, a feature not seen on the standard Model C.

Case Model CI (industrial) tractor fitted with industrial rubber tires and a cable-operated "fabri-form" loader shovel, cleaning a highway shoulder in Missouri. Photo dated 1931. *Case Archives*

by Oliver. Mechanical power implement lifts were introduced in 1928 with the Deere GP. The Farmall F-12 had one in 1933. The Deere Models A and B sported hydraulic power lifts in 1934 and 1935, and the 1935 Case CC featured the new "Motor-Lift."

In 1935, Oliver introduced the Row Crop 70 tractor. It brought about one of the most profound turning points in the farm implement industry's direction. It reflected both the optimism of the improving economic situation and the increasing influence of the automobile on tractor design. In 1933, 1934, and 1935, durable goods companies were beginning to see that styling and product differentiation had a positive effect on sales. Up to the 1930s, cars, for example, were quite a bit alike: boxy, four cylinders, clam fenders, exposed radiators. By the mid-1930s, styling was individualistic. Six-, eight-,

Moline, and even Sears-Roebuck, with their Bradley GP.

All of the above were smaller, lighter, and cheaper than the Case CC, with the exception of the Farmall F-30. Yet Leon Clausen resisted the urging by his sales force for a smaller version of the CC, saying, "The purpose of Sales is to sell what you've got!" Finally, Clausen yielded. The Model RC, rated for 11 drawbar and 17 belt horsepower, was introduced in 1935.

By 1935, the effects of the Depression were diminishing, as far as the tractor business was concerned. Industry sales for 1935 were up to over 160,000 units. Changes thus far in the 1930s included the advent of diesels, pioneered by Caterpillar; rubber tires, pioneered by Allis-Chalmers; and high-compression engines powered by gasoline, instituted

Model CC and CC4 tractors were usually sold with fenders. At least three different fender and bracket designs were sold during this model's production run.

twelve-, and even sixteen-cylinder engines had replaced the fours, and radiator grills were the hallmark of styling.

Although there had been some attempts at tractor styling before, the new Oliver 70 immediately overshadowed the competition. It influenced tractor design from then on. It was styled. It was powered by a six-cylinder engine. It could be equipped with an electric starter and lights. Its high-compression engine was designed to run on 70 (hence the model designation) octane gasoline. It had an instrument panel and finger-tip controls.

The Oliver 70 (now with a big "Oliver" and a smaller "Hart-Parr") was available in four configurations: row crop, standard tread, orchard, and high crop. Each configuration offered the option of the HC (high-compression) gasoline engine, or the KD (kerosene-distillate) engine. It was also sold in Canada as the Cockshutt 70.

Several third-party vendors produced crawler-type conversions for agricultural tractors during the 1930s. Popular among these firms were Roadless and Trackson. Pictured above is Model C with Trackson conversion unit pulling a Fresno scraper. Steering was accomplished with only the differential turning brakes, which meant the opposite track's speed increased on turns. *Case Archives*

The Model CC row-crop tractor could be fitted with a wide front axle, which made a CC4. A few units produced by Case in this (wide axle) configuration were serialized CH rather than CC. There are those who point out that these units had no turning brakes as did the CC and CC4. Owner: Scott Fuller, Clinton, Iowa.

Model RC

In 1932, International Harvester announced its smaller Model F-12 general purpose tractor. This tractor not only was able to replace the horses on small farms, but also was finding considerable use on large farms as a back-up tractor. Tractor makers were realizing that the small farmer simply could not afford the bigger tractors, and that large farmers were buying several tractors of various sizes.

By 1935, several more manufacturers had entered the under 20-horsepower market. John Deere was ready with their Model B, and Case was ready with their Model RC, which was a one-plow row-crop general purpose tractor.

Another example of a Model C with a trackless conversion unit. Visible below and to the left of the steering wheel is one side of the linkage that operated the differential turning brakes.

The RC featured a four-cylinder engine supplied to Case by Waukesha. It displaced 132 cubic inches and was configured for gasoline only. Its rated speed was 1,425 rpm. The engine was equipped with a governor and a conventional radiator and fan. Rather than a water pump, however, the thermosyphon system was used.

The tractor was introduced in the row-crop general purpose configuration. The rear tread was adjustable from 44 to 80 inches by sliding the wheels on the splines and by reversing the wheels. A PTO was standard, as were steering brakes. An adjustable wide-front was an option, so were rubber tires. A three-speed transmission was provided. Overhead steering was initially furnished, but the steering ratio proved to be too high for the usual Case feel that came with the "chicken-roost" side arm steering. By 1937, the chicken roost had been adapted.

In 1937, Case bought the Rock Island Plow Company. Since the facility was empty, production of the RC was transferred there immediately. In 1938, the company introduced a standard tread version: the Model R.

By 1939, the R and RC were each given a cast iron "sunburst" grille and the new Flambeau Red paint. They also got a four-speed transmission and a

Shown above is a correct example of the C series later production. Zenith carburetor and V-type fan belt became standard during the early part of the decade. Side view of this CH (identical to CC4 except for brakes) shows ample crop clearance for cultivation.

The Model L engine was built with only minor changes from 1929 until 1953 (in the Model LA). Shown in this photo is a 1929 Model L with the later Zenith carburetor. The Model L pioneered a patented roller chain final drive from the differential to rear axles, designed by David P. Davies, Case engineer. This design lasted into the late 1960s. Owner: Jay Foxworthy, Washburn, Illinois.

Rear view of the CH model shows drawbar height and cast steering wheel, typical of all C series tractors.

CASE

PREVIOUS PAGES
The Model L became famous for its sterling performance on the belt and, in threshing and sawmill applications, where it consistently outperformed large steam engines with its steady, dependable power.

The General Purpose Tractor

As of 1927, the Fordson was gone, at least from Detroit. Ford had seen the changing tastes in tractors by American farmers and had transferred production to Cork, Ireland. Irish-made Fordsons would be imported over the next 12 years, but their competition would be nothing like it was in the 1920s. Henry Ford told the public that he needed the Fordson's manufacturing space to make the forthcoming Model A car.

The farmers' changing taste in tractors was mainly due to the new offering from International Harvester: the Farmall. The Farmall, introduced by International Harvester in 1924, unveiled a totally new era in the tractor business. No longer was the steam engine the competition—the tractor had taken on the horse. From now on, the tractor would be required to do more than just pull with the drawbar or power with its belt pulley. It would be required to power implements (such as binders and mowers with power take-off (PTO) shafts) and mount integral cultivators, mowers, and plows, which could be lifted through engine power. Wheel spacing would have to be adjustable for differing row widths, crop clearance would have to be greater for cultivating corn, etc., and maneuverability would have to be greatly improved. The age of the "all-purpose" tractor had arrived.

Jay Foxworthy, of Washburn, Illinois, bought this 1929 Model L at an antique auction for a mere $225. Needless to say, Jay has invested considerable time and money in it since then.

The Case Model RC appeared in 1935 and was built through 1940. Shown here is a 1939 model painted Flambeau Red and sporting the "chicken-roost" side arm steering shaft.

A 1939 advertisement features the new A-Six combine. It's being pulled and powered by a Case Model CC tractor.

starter and lights option. The RC got a Motor-Lift.

The birth of the RC was not an easy one. By 1934, International Harvester dealers were eating the Case salespeople's lunch with their smaller F-12 Farmall. Branch people and the Case sales manager sent letters to President Clausen urging the rapid development of a Case counterpunch. Clausen resisted, saying a smaller Case would just take sales from the larger, more profitable Model CC. He also pointed out that new implements would have to be developed, which would detract from other engineering efforts going on at the time. Finally Clausen acceded on the condition that the new small

The Model L was billed by Case as the most tested tractor ever built. More than 15 prototypes were tested in farming conditions the year before the L was formally introduced.

LEFT
The engine of the Model L has a 4.625-inch bore and a 6-inch stroke. The original flat fan belt on this 1929 model was replaced by a Case conversion kit to a V-belt.

BELOW
The Model L was an unexcelled plowing tractor. In most conditions, it handled four 14-inch bottoms with ease. More than 30,000 Model Ls were sold during its 10-year production life.

Standard-type Model R in styled configuration. For reasons that remain unknown, almost all of these later production model series were shipped to Canada. Owner: John Davis, Maplewood, Ohio.

tractor, to be called the RC, would be, in his words, "An orphan, so as not to cut into Model L and C sales." The RC was to be painted a lighter gray for differentiation, and there was to be no advertising hoopla, only a brochure. Clausen also cautioned all concerned that the size and power of the RC were to be carefully chosen; under no circumstances would they be changed.

Clausen continued to worry and drag his feet on the RC project, which he called a "half-tractor." He was afraid the "weak-kneed" salespeople would sell it for jobs over its head, rather than the more expensive Model C or Model L. In January 1935, Engineer Davies asked Clausen if a fourth gear should be added to the RC's transmission for an extra cost of 72 cents. Despite the fact that the new Deere and Allis-Chalmers small tractors had four speeds, Clausen said no. He cut down on cost wherever possible, and his final instructions were, "Leave off the fenders." Fenders were also optional on the John Deere and the Farmall.

An example of the styled RC, with "sunburst" cast grille. Note the traditional "chicken-roast" steering linkage on left side. Owner: Loren Engle, Abilene, Kansas.

Waukesha engines powered all R series units. The Model R used a 132-cubic-inch engine rated to run at 1,425 rpm. Note the Case-supplied oil pan with hand holes, characteristic of all Case units of the time.

The Model R in standard configuration. The powerplant and drivetrain are identical to the RC. The styling of this Model R indicates it was built in 1938 or later. Note the cast grille and scalloped hood design. Later production (styled) R series had four-speed transmissions rather than the original three-speed unit used in earlier Model Rs. *Case Archives*

Right-hand view of the styled Model RC. Late production examples of this model usually had the electric start option, with the battery located on top of the steering column, behind the fuel tank.

CASE
CASE
CASE
CASE
IMPORTANT
KEEP COVER TIGHT ON GASKET
TO PREVENT DIRT or WATER
FROM ENTERING MOTOR

Chapter 5

The Flambeau Red Era, 1939–55

Case tractors for 1939 sported a new paint color: Flambeau Red, so named because "flambeau" is the French word for flame or flaming torch. Also, the Flambeau River region of Northern Wisconsin was the original home of the eagle, Old Abe. The new color was an orange-tinted red, unfortunately close to Allis-Chalmers' Persian Orange. Nevertheless, the color, not to mention the new tractors, were attractive and striking.

The Flambeau series consisted of Models D, S, V, and LA, replacing the previous gray Models R, RC, CC, and L models (1939 Model Rs were given the new Flambeau paint, however). They sold well. Case built 390,000 Flambeau Red tractors by the end of 1955, more than twice the number of internal combustion tractors built previously. Ford,

LEFT
The SC model pictured above in row-crop configuration was the most popular of the S series. The S engine was the highest rpm powerplant in the Case line and typified the shorter stroke, higher rpm units that would become popular in the 1950s.

The Case S series was an all-new two-plow tractor design brought into the marketplace in late 1940 to stand model-for-model against competitive brands. Owner: Elwood Voss, Ashton, Illinois.

First-year production S series did not have the horizontal stiffening ribs along the lower block sides. Cylinder bore of 3.5 inches was increased to 3.625 on last-year production units.

however, built 950,000 during that time, making them second behind International Harvester. Deere was third, and Case was fourth during this period.

D Series

The first of the new styled Flambeau tractors, the Model D Series, replaced the previous C Series in July 1939. The Model D was not actually a completely new machine, however, but basically a styled Model C. The engine, drivetrain, brakes, and steering were initially similar to that of the Model C. At first, there was not going to be a model letter change, but after 500 or so were built, the D designation was started. Industrial tractors retained the Model C designation through most of the model year, however.

The D was the standard tread version of the series. The DC was the general purpose version. It was offered as the Model DC-3 row crop, DH hi-clear, and DC-4 solid wide axle. The DO and DV were orchard and vineyard versions. The industrial version was the DI. There were several other minor variations as well.

For 1940, the three-speed transmission was replaced with a four-speed unit. The DC-3, with Case's traditional "chicken roost" steering, was initially

Yield-Per-Man is the pitch of this 1948 *Successful Farming* advertisement.

available in the tricycle configuration only, with either single or dual front wheels, either on steel or rubber. Later, an adjustable wide-front was offered as an option. The DC-4 had a standard front axle, like the D. The DH, which preceded the DC-4, was the same as the DC-4, except it did not have steering brakes as did the DC-4.

The DC-3 and DC-4 had a standard PTO, steering brakes, and the Motor-Lift. All Ds had swinging drawbars, fenders, belt pulleys, and water temperature and oil pressure gauges. Options included a muffler and rubber tires.

Over the years, the D was upgraded considerably. The hand clutch gave way to a foot clutch, hydraulics replaced the mechanical Motor-Lift, disk brakes replaced the old band-type, and a factory LPG (Liquefied Petroleum Gas) version was offered. A thermostat and water pump were added in later years, as were live PTO and hydraulics. Finally, in 1952, the Case Eagle Hitch was added. Production ended in 1953.

S Series

In 1939, when the Flambeau Red Case line was introduced, the two- and three-plow general purpose tractors were the largest sellers. John Deere had just styled their Models A (two-plow) and G (three-plow), which were selling very well. International Harvester, for 1939, had come out with a styled tractor line. Featured were two look-a-likes, the Farmall H and the Farmall M. The H was a two-plow tractor; the M was a three-plow. Oliver competed with their Models 70 and 80. Standard tread counterparts and orchard and industrial versions were available as well as the general purpose types. Not to be outdone, Case offered a pair of similar appearing tractors: the Models S (two-plow) and D (three-plow).

The unit above is a 1942 unit, of prehydraulic production. Note the mechanical motor-lift unit at back of the differential, below operator's seat.

Another example of the SC model. Later production units saw improvements like hydraulics (single and dual spool) and the change from hand to foot clutch. Owner: Kelly Clevinger, Indianola, Iowa.

The Case Model S was announced in November 1940. Unlike the D, it was all new from top to bottom. It was available in the standard tread version, the plain Model S; the general purpose version, known as the SC; the SO-Orchard version; the SI-Industrial version; and the SC-4, built only in 1953 and 1954 with a fixed-tread wide front axle. Production continued until June 1954.

The S Series tractors were equipped with a relatively short-stroke high-speed engine. This series began life with a 3.5-inch bore and 4-inch stroke. Rated speed was 1,550 rpm. In 1953, the bore was increased to 3.625 inches, and the rated speed was raised to 1,600 rpm. A four-speed gearbox was provided.

This 1941 *Successful Farming* ad compares the Case SC to a good gun or horse. Early Model SCs are easily identified by the exhaust that exited on the hood centerline, rather than along the side.

V Series

By 1939, it was clear there was an expanding market for the smaller one-plow tractor. Deere and Allis-Chalmers had pioneered the field in 1937. The John Deere Model L, which was about the size of today's garden tractors, weighed only 1,500 pounds, had about 10 horsepower, and cost $450. The Allis-Chalmers Model B weighed about 2,000 pounds. It was more powerful, at 15 belt horsepower on kerosene. The Allis-Chalmers B sold for $495. Deere also came out with the Model H in 1939, which was a full general purpose row-crop tractor, but about three-fourths the size of the Deere Model B. International

A high-clearance version of the VA series, this VAH was built in 1953. Special purpose units of this sort were popular in cane-raising applications for cultivation in areas like Louisiana. Owner: Weston Link, Ridgeland, Wisconsin.

The VA series WH tractors used a bull-type reduction gear on the rear axles and 24-inch rear tires. By using this reduction gear configuration with 36- or 38-inch rear rubber, Case engineers were able to produce the VAH high-clearance model with the addition of special front axle spindles.

Harvester had new, capable small tractors in the Farmall Models A and B. Prices were less than $700.

Case, specifically President Clausen, had trouble seeing the point of a light, small tractor. This was due, in large part, to the fact that the Models R and RC did not sell well. In fact, in 1939 there was still a considerable stock of new, unsold Rs and RCs in inventory. Nevertheless, Case salespeople pressured management not to abandon such a lucrative market segment.

In June of 1939, Clausen finally relented and gave the go-ahead for a 15- to 17-horsepower machine (belt) in the 3,000-pound class to be built in the Rock Island plant for the 1940 model year. It was to be identified as the Model V; row-crop versions would be the Model VC. Also to be offered were the VO (orchard), and the VI (industrial).

The V was developed to use a Continental four-cylinder engine of 124 cubic inches. It was equipped with a four-speed transmission and a gear final drive with gears supplied by Clark Equipment Company. In typical Case fashion, the engineers overshot on both power and weight. The V was rated at 22.07 horsepower on the belt at the University of Nebraska. Its test weight was 4,290 pounds. Accordingly, maximum drawbar pull was 2,798 pounds. Under similar circumstances, the Allis B pulled 2,667 pounds; the Farmall A, 2,387 pounds; the Ford-Ferguson (without the load-transferring benefits of the three-point hitch) made 2,236 pounds; and the John Deere H pulled 1,839 pounds (on distillate).

Interestingly, Case's mid-sized new tractor, the Model S (also brought out in 1940), had a rated belt horsepower of 21.62, slightly less than the small-sized Model V. Maximum drawbar pull of the S was higher, however, at 3,166 pounds. This disparity would have been greater if Clausen had not stepped in and insisted that the Model V's drawbar power be rated at 1,425 rpm, rather than 1,650 rpm. Since the

This 1941 *Farm Journal* ad shows the price for the VC at $630.

An example of first-year production D series, this DC has the correct early production deep-dish fenders, low dashboard, and older C-style oil pressure indicator. The first 500 D series production units were Model C units changed over to the new styling. Owner: Warren Kemper, Wapello, Iowa.

V was a Rock Island tractor and the S was built in Racine, competitive feelings existed. The "old-timers" in Racine did not want the Rock Island newcomers upstaging them.

Production of the V series ended in 1942, being replaced by the VA series.

Three-Point Hitch

"You haven't got enough money to buy my patents," Harry Ferguson bluntly told Henry Ford, the richest man in the world in 1938.

"Well, you need me as much as I need you," Ford reportedly responded. "So what do you propose?"

"A gentleman's agreement," retorted Ferguson. "You stake your resources and reputation on this idea, I stake a lifetime of design and invention—no written agreement could be worthy of what this represents. If you trust me, I'll trust you."

"It's a good idea," said Ford. With that, both men stood up from the kitchen table that had been placed outside for their meeting and shook hands.

Thus was born the Ford Tractor-Ferguson System, or Ford-Ferguson, Model 9N. It was one of the five developments that revolutionized power farming and decidedly affected Case's fortunes. The five developments were the inexpensive Fordson of 1918; the general purpose Farmall of 1924; the car-like Hart-Parr Oliver 70 of 1935; the Ford-Ferguson with its hydraulic three-point hitch of 1939; and the Farmall MD of 1941, which normalized farm diesel power.

Irish inventor-entrepreneur-salesperson Harry Ferguson (1884–1960) had patented a mounted plow for the Fordson back in 1924. The plow featured two hitch points; one above and one below the line of draft. The idea of the upper and lower links was to prevent Fordson backflips. Ferguson's plow solved that difficulty. The problem was that when the front wheels ran up on a

This 1953 ad touts live hydraulics and live PTO for the Models SC and DC tractors.

Case's Model 500 weighed over five tons in working condition. The roller chain final drive featured in 1929 Model Ls was still used on the 500 and would continue to be used for another decade. During the Nebraska test of 1953, the 500 demonstrated 64 belt horsepower and a drawbar pull of 7,400 pounds.

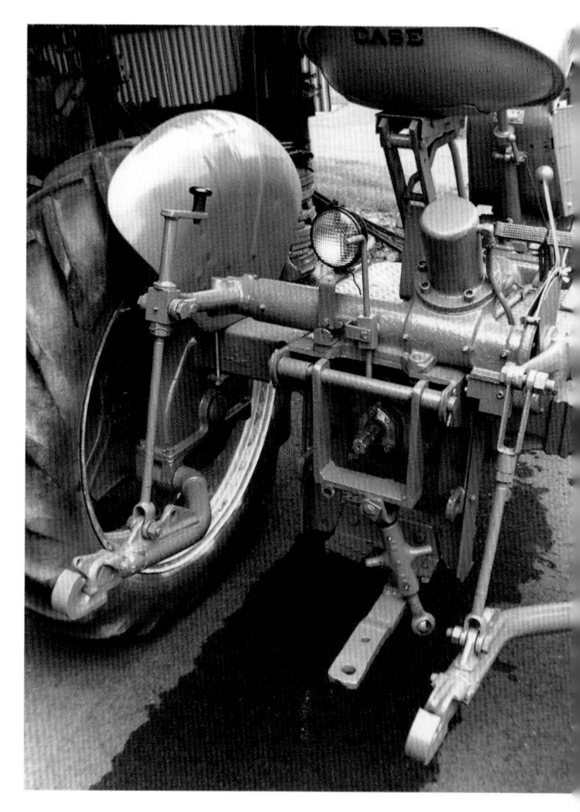

Case's answer to the three-point hitch was the Eagle Hitch system, first put into production in 1948 on VAC units like this shown. Draft control was nonexistent, but the hitch system was a help to most farmers.

hump, the plow went deeper. Depending on how it was adjusted, it was possible for the plow to continue going deeper even after the hump was past. Naturally, the tractor would bog down and stall. Conversely, if the front wheels dropped into a ditch, the plow would be lifted from the ground. Ferguson overcame these difficulties, after a fashion, by using a semirigid mounting with springs and a depth-gauge skid plate behind the plow.

Ferguson and his team recognized that hydraulics were the answer to this problem and so rigged up a Fordson with an add-on system. They introduced a load sensor into the bottom link. As draft loads increased, this sensor would signal the hydraulics to lift the plow until the load returned to normal, and conversely, with decreasing loads. This didn't work too well, be-

DIESEL
"500"
CAS

PREVIOUS PAGES
The Model 500 was essentially a diesel version of the LA model. It was Case's first true diesel offering in an agricultural tractor, and production was begun in 1953. While L and LA units were produced earlier with the Hesselman diesel conversion to Case's own gas engine, this six-cylinder engine was a true diesel and performed well. The engine was of the Lavona power-cell patent design, with open chamber combustion system. Owner: Clyde Barrows, Rochelle, Illinois.

cause the loads on the lower link were too severe. Using the upper link for sensing worked better.

At this point, Ferguson discovered the concept of two nonparallel lower links and a single sensing upper link, all three pointing at a point slightly ahead of the tractor. This caused the virtual, or apparent, hitch point to be at that locus. With uniball attachments at both ends of each link, the required degree of flexibility was provided to allow for steering with a plow in the ground. Also, it made a rear mounted cultivator practical. It prevented the cultivator from swinging in the opposite direction when the steering wheel was turned.

In this concept, the upper link pushed forward on a hydraulic control valve in proportion to the rearward, or draft, load on the implement. Once the hand-operated hydraulic control posi-

Seven main bearings on the Model 500's six-cylinder 377-cubic-inch engine helped make this one of the strongest designs and most powerful diesel engines offered to farmers in the early 1950s.

tioned the implement to the desired depth setting, the upper link manipulated the hydraulics, raising or lowering the implement as required to keep draft load constant. When set up properly, the three-point hitch made plowing an absolute snap. When the Ford-Ferguson was first introduced to the press in 1939, a plowing demonstration was given by an eight-year-old boy. Even in small fields, the 120-cubic-inch, 2,500-pound $595 Ford could plow an acre per hour.

Ferguson had an abortive arrangement with David Brown before meeting with Ford in 1939. A few tractors were built but did not sell well because of price. Ferguson and Brown decided to split (although it put Brown in the tractor business).

In June 1939, a month before Case revealed the first of the Flambeau Red tractors, the Ford-Ferguson was introduced to the press. Despite introductory manufacturing and distribution problems, more than 10,000 of the little gray tractors were sold that year.

What made the Ford-Ferguson such an instant success? It was Harry Ferguson's draft-load compensating three-point hitch and his customized mounted implements. The little 2,500-pound Ford-Ferguson, which sold for $595, could plow as many acres in a day as tractors weighing and costing double. It was quiet, safe, easy to mount and dismount, and car-like to handle. A single version, which came to be known as the utility configuration, served as a general purpose tractor, plowing tractor, orchard tractor, and industrial tractor.

Despite the onset of World War II with its production and material limitations, almost 200,000 Ford-Fergusons were sold in the next seven years before Ford and Ferguson went their separate ways. Most of the traditional tractor makers did not take the Ford-Ferguson seriously. During 1940, the first full year of production, more than 35,000

The Case Low-Seater is the feature of this 1953 *Farm Journal* ad. The Low-Seater was properly called the VAC-14. It was built to resemble the highly-successful Ford and Fordson utility tractors. The VAC-14 had a low seat with the operator straddling the transmission, a low steering wheel, a utility-type front axle, and starter, lights, and muffler as standard equipment, as did the hydraulic Eagle Hitch. A wide adjustable front axle was offered by Case for both the SC and DC tractors. Adaptable for crop widths of varying ranges, this configuration was popular in specialized crop applications.

Case tractors are becoming increasingly popular at steam power shows throughout the United States. This restored SC is pulling a PTO-driven binder. Owner: Elwood Voss, Ashton, Illinois.

of the little gray tractors were sold. Some of the long-line companies began to react, but World War II intervened, preventing a response until later. Case legend has it that Case President Leon Clausen had met Ferguson on a steamship prior to Ferguson's meeting with Ford. Ferguson supposedly offered the three-point system to Clausen, but Clausen was not interested, saying that real tractors had enough weight and did not need such a system.

When Young Henry Ford took over from his grandfather in 1946, he soon recognized that Harry Ferguson was making all the money on their handshake agreement. He therefore told Ferguson that as of mid-1947, the agreement was dissolved. With that, Ford set about revising and improving the tractor from the Ford-Ferguson into the Ford 8N (the "8" signifying model year 1948). He set up his own dealer network (previously, it had been Ferguson's), and built a complete line of three-point hitch implements (these, too, had previously been Ferguson's). Ferguson countered by launching his

The S Series tractors were equipped with a relatively short-stroke high-speed engine. Early versions were rated to run at 1,550 rpm. In 1953, the rated speed was raised to 1,600 rpm.

CASE
CASE

While adjustable for a wide range of crop widths, the wide adjustable front end was not nearly as practical as the row-crop version. While fine for specialized applications, few of these configurations got many hours on them on typical farms.

The Model VAC is featured on this *Successful Farming* ad of 1951.

own U.S. tractor, the Ferguson TO-20, importing his TE-20 (already in production in England) until it was ready, and suing Ford for loss of business and patent infringement.

World War II Production

The sneak attack on Pearl Harbor was a shock to most Americans when it happened on December 7, 1941. At American industries such as Case, however, it was common knowledge that war was imminent. Already in September 1940, Case bid for a contract, which they were subsequently awarded, to make 155-millimeter howitzer shells. Over 1.3 million were made in the next five years. Most of Case's wartime effort was in subcontracting items to specialized military suppliers. For example, Case built about 185 pairs of wings for the Martin B-26 Medium Bomber and many aftercoolers for the Packard-Rolls-Royce Merlin aircraft engine. Case also supplied 15,000 Case tractors, of 11 models, to the military during the war. One specialized version of the Case LAI was adapted for desert duty. It boasted air brakes, folding windshield, and blackout lights, and had a top speed of over 40 miles per hour.

One wartime development that hurt Case was the appliance of government price and production controls. The government faced two problems: build

Case V and VC models followed the R series in production at Case's Rock Island, Illinois, plant. Built primarily from vendor parts, the V series used a Continental L-head engine, Clark transmission, and vendor-purchased sheet metal.

war materials and produce food. Critical materials, such as steel and rubber had to be divided between these two demands. To solve this dilemma in the agricultural equipment industry, the government decreed 1942 production for farming would be at 83 percent of the material volume used in 1940. Products introduced after 1940 could not be built at all. This excluded most of the V Series and S Series. Although relief was obtained, Case fell far behind its competitors in product development during the war.

Although the full extent can only be estimated now, it is known that Henry Ford had good connections in the government and was one of the biggest producers of war materials. It was rumored that he made an effort to get all tractor raw materials allocated to himself, saying he, and his little Ford-Ferguson, could make the best use of it for America. It is also known that Charles Wiman, president of Deere & Company, temporarily resigned his post at Deere to become a colonel in the military procurement operations. Leon Clausen, on the other hand, had been an outspoken opponent of government involvement in private business. (He had gone so far as to declare he would not participate in the Social Security System.) How much these government relationships helped or hurt wartime tractor quotas is now moot. Case was not in an enviable position. Nevertheless, about 15,000 of the V Series (mostly VCs) were sold before the model was terminated in 1942 in favor of the VA Series.

VA Series

The Rock Island Model V was a good tractor. Excluding the advantage of the Ford-Ferguson's three-point hitch, the V and VC were probably the

Shown above with steel wheels, VCs of this configuration were usually built for export. Note chrome trim on hood, which designates early production years for all Case Flambeau Red series units.

YOU CAN GET ALL THESE AND MANY MORE IMPLEMENTS FOR THE

CASE "VAC"

LOW-COST TRACTOR

LATCH-ON IMPLEMENTS

THE ONLY TRACTOR WITH ONE MINUTE HOOK-UP TO

LATCH-ON IMPLEMENTS

FROM TRACTOR SEAT

LATCH ON... SLIP IN PIN AND GO

"NO FINER AND EASIER IMPLEMENT HOOK-UP ON THE MARKET"

"Farmers around here like the Case 3-point Eagle Hitch. I think there is no finer and easier implement hook-up on the market. The Latch-On implements are unexcelled for sturdiness and quality." — Walter Dyrland.

MASTER-FRAME IMPLEMENTS

"FOR MY MONEY THE GREATEST BUY IN TRACTORS"

. . . says M. V. Harris. "My Model 'VAC' has given perfect satisfaction in every kind of work. It requires very little service, and fuel costs are amazingly low on all farming operations." See for yourself how easy the Eagle Hitch latches onto implements . . . how it pulls plows at even depth in uneven ground . . . how its balanced down-pull helps *both* steering and traction. Compare its implements with those for any other low-cost tractor — for handy hydraulic control, for uniform penetration. Besides the great variety of Case implements, there are loaders, hole diggers, etc., built by specialty manufacturers for the "VAC." Plan now for the work ahead . . . see your Case dealer.

CASE

21 great tractors in four power sizes and many models offer you exactly the tractor to fit your acreage and crop system. Get full information from your Case dealer.

PASTE ON PENNY POST CARD AND MAIL

Name
Postoffice
RFD State

SUCCESSFUL FARMING, JANUARY, 1951 57

Notice this 1951 *Successful Farming* advertisement calls for pasting the coupon on a penny postcard. How times have changed!

best of all the new less-than-$700 tractors. They had contemporary styling, good power, and good weight distribution. What they lacked could be made up for by the addition of wheel weights. Nevertheless, Case was squeezed from a profit standpoint. To raise the price would make it competitive with the Model S and make the S superfluous. The fact that the engine, gears, and some sheet metal for the V came from outside sources was cutting into the margin, however. Also, the Continental flat- head engine could not be made to perform well on distillate, which was still favored by the farmers.

Thus, a completely new tractor was designed to replace the V: the VA. General purpose versions were called the VAC. While they looked much the same, not much was interchangeable. The new tractor used a Case-built overhead valve engine, still with 124 cubic inches of displacement. At first, Case supplied block castings to Continental for finishing, but when the Rock Island

During the early 1950s, Eagle Hitch was offered on Case's SC and DC series tractors. While this hitch system did not offer true draft control, it became immediately popular for its ease of use.

engine plant was finished, the engines were all Case.

The VAC, like the VC, used the chicken-roost steering system, which was a favorite of Clausen's after the cost-saving attempt to use overhead steering on the RC. By 1946, the VAC had gone to a modified drag-link-type of steering. In 1951, a contemporary universal-joint type was substituted.

The VA Series was the first to employ Case's answer to the Ford-Ferguson three-point hitch: the hydraulic Eagle Hitch. Introduced in 1953, the VAC-14 was built to resemble the then-separated Ford and Ferguson tractors. It had a low seat with the operator straddling the transmission, a low steering wheel, a utility-type front axle, and starter, lights, and muffler as standard equipment, as did the hydraulic Eagle Hitch. It must have anguished Leon Clausen, who was then Board Chairman.

The V Series was made from 1942 to 1955. Almost 60,000 were sold, of which some 15,000 were VAI industrial models.

Eagle Hitch

In 1948, Case introduced its version of Ferguson's three-point hitch, which the company called the Eagle Hitch. It had the advantage of rapid snap-type attach points, but lacked Ferguson's patented "Draft Control." Management of the big tractor companies, Case included, thought the three-point hitch was only viable for the smallest tractors. To Case President Clausen, who was never a fan of lightweight tractors anyway, the Ferguson system was a crutch; "a cheatin' system," he called it. Nevertheless, farmers liked it, despite the fact they had to buy new implements. Once the farmers had some three-point implements, they were not likely to buy a tractor that could not use them.

The year 1947 saw Ford and Ferguson split. Now there were two three-point tractors available. Also in 1947, Deere came out with its Model M, which had a nondraft load compensating three-point system that could use Ferguson implements. TerraTrac, the new crawler outfit from Churubusco, Indiana (which we will hear more of later) was the first to simply adopt the Ferguson system in 1950. Oliver introduced the Super 55 utility tractor in 1954. It not only had the draft-load three-point hitch, but also looked like the Ford and Ferguson tractors. International Harvester, like Case, was not inclined to acquiesce. In 1954, International Harvester announced their two-point Fast Hitch. While it worked well, it could not use three-point implements, and vice versa. Finally, in 1958, they offered the option of two-point or three-point systems. By 1960, a draft-load compensating three-point hitch was universally accepted for virtually all tractors.

Case's Model D was the standard version of the more common DC row-crop tractor. Both units shared common drivetrain components. Smaller wheel size of the D gave it slightly less drawbar pull than the larger-wheeled DC. Owner: J.R. Gyger, Lebanon, Indiana.

Model LA

The flagship of the Flambeau Red era, the LA, was merely a styling update of the long-lived Model L and came out in 1940. The LA was updated by an engine compression ratio increase, some engine oiling improvements, the addition of the four-speed transmission of the old LI, and a PTO and electrical system as options. The Model L was introduced in 1929 to compete with Deere's successful Model D. The Deere D had a 30-year production run, a record for the tractor industry. Case's Model L and LA Series enjoyed a 24-year history.

This SC is shown in plowing form, pulling a Case Centennial two-bottom plow. The SC pulled two plows well in most soils. Owner: J.R. Gyger, Lebanon, Indiana.

This restored Model VAH is typical of the improved serviceability of newer Case equipment of the early 1950s. Note live engine-driven hydraulic pump and spin-on oil filter. Owner: Weston Link, Ridgeland, Wisconsin.

The big Model LA was offered only as a standard tread version. It was offered with steel wheels as late as 1949, although rubber was by far the most common. It was offered in gasoline, distillate, and LPG (liquefied petroleum gas) versions, although gasoline was by then the most common. The gasoline version made 57 belt horsepower during its Nebraska test; the distillate version, with lower compression, made 47 horsepower; the LPG version, with a compression ratio higher than that of the gasoline version, made an astounding 60 horsepower. Not too bad for a 400-cubic-inch engine at 1,100 rpm.

Model 500

The year 1952 was not really the end of the line for the venerable Model LA. It was resurrected the following year as the Model 500, the first of the new three-number tractors. The big news was that the 500 was a diesel, a first for Case.

The 500 was originally identified at Case as the LAD (LA-diesel). The tractor was essentially identical to the LA with the exception of a higher axle ratio due to the higher speed of the diesel. Even the hand clutch was retained. The bare weight of the tractor was about 7,500 pounds, although working weight could approach 10,000 pounds. The 500 was able to develop 7,400 pounds of drawbar pull and 64 belt horsepower during its Nebraska test.

RIGHT
The VA Series replaced the V series in 1941 when Case built their own engine to use in place of the Continental engine of the V Series. Shown here is a rare VAS model, of which only about 1,600 were built. The VAS was an offset high-clearance row-crop tractor.

BELOW
D series tractors were made in industrial and orchard versions. The Model DO pictured above is in full orchard dress, or "frosting" as termed by orchard owners. Owner: J.R. Gyger, Lebanon, Indiana.

CASE

This nicely restored Case VAO is a 1947 model that did its work in the orange groves of California before retiring. Owner: Dan Buckert, Hamilton, Illinois.

A 1944 standard tread Model S. The S was promoted as a smaller version of the D, in a manner similar to that of the John Deere Models A and B. Owner: J.R. Gyger, Lebanon, Indiana.

The 377-cubic-inch six-cylinder engine was all new. It featured a main bearing between each connecting rod. The engine was of the indirect injection type that employed the Lavona "power-cell" combustion chamber. This was the same type used by Oliver in their 88 and 99 diesels and by Mack trucks.

The Model 500 was built from May 1953 through April 1956. It was the last of the all Flambeau Red tractors.

Summary of the Flambeau Years

By 1952, tractors finally outnumbered horses on American farms. Although the shortages of the war years compounded post-war demand for trac-

A VAI (industrial) version of Case's VA series. This one was built for military applications and has a Detroit mower attached. Owner: J.R. Gyger, Lebanon, Indiana.

tors, Case was falling behind in getting its share of the business. Much of the blame for this fell on President Leon Clausen and his "we'll tell you what you need" marketing attitude. Clausen's attitude toward organized labor also led to a devastating strike after World War II. During the strike, many of Case's best dealers were lost.

Because of World War II activities, Case also lagged in tractor development. Deere, International, and Oliver all had diesels long before Case. Ford had introduced the three-point hitch in 1939; the Case Eagle Hitch was not out until 1949. International IH tractors had five-speed transmissions; John Deere and Oliver had six, while Case tractors only

CASE
CASE

This is a Model SI (industrial). Late production units of this type were designated Model 30s and usually had a loader attached. They were built for work in Florida citrus groves.

had four. Case tractors were incredibly durable and generally more powerful than their direct competition, but they were not upgraded over the years. Especially after the surge of post-war production was over, these things hurt sales. By then, none of the Flambeau tractors were a market leader.

In 1946, Case commissioned an eye-opening survey of farmers. The farmers indicated they did not like the hand clutch; they wanted more forward speeds (Clausen believed tractor speeds over 12 miles per hour were dangerous), diesel engines, and the frills and styling that Clausen abhorred. Case's styling was ranked sixth behind John Deere, International, Ford, Massey, and Oliver. Farmers objected to the Flambeau Red color, saying it did not hold up, and they did not like the chicken-roost steering, regardless of the improved controllability.

To the above there was only a causal response by a lackluster Board of Directors. Theodore Johnson replaced Clausen as president when Clausen became Board Chairman. Johnson, and indeed all the Board, were in their late sixties.

In 1953, Johnson was replaced by 54-year-old John T. Brown. Board members were also being replaced by younger men, and in 1955, Clausen resigned as Chairman. This paved the way for the new three-number tractor series.

The Model LA was Case's largest tractor in the Flambeau Red lineup. Available in either gas or LP version, this model is a direct descendant of the original 1929 Model L. *Case Archives*

CASE
Farm Machinery
Est.1842

CASE
400

Chapter 6

Tractors of the Desert Sunset, 1956–76

When John T. Brown became president of Case in 1953 he brought onto the Case Board some energetic younger members. One of these was William J. Grede. To streamline the decision-making process, an Executive Committee was formed, chaired by Grede. By 1954, Case introduced 22 new products. Two of the most important, for model year 1955, were the new Case 400 tractor from Racine and the 300 from Rock Island, both in striking new Desert Sunset-painted sheet metal with Flambeau Red cast iron and wheels. Along with the 500 diesel, brought out in 1953, these modern and capable tractors replaced the entire line.

Model 400

Since the Model 500 was strictly a standard tread tractor, a large, general purpose tractor was needed. The Model 400 filled that bill. The 400 was Case's first totally new tractor since 1928.

LEFT
The Model 400 replaced the D series in 1955. Available in three fuel varieties and many front end configurations, this single front variety is seldom seen. Owner: J.R. Gyger, Lebanon, Indiana.

Mark B. Rojtman (pronounced "Roitman") was president of the American Tractor Company when that firm was purchased by Case. He was made a Case vice president and shortly thereafter became Case president in 1957. This man was responsible for the Case-O-Matic transmission and for the headlights-in-the-grille styling that Case adopted for their entire tractor line in 1958.

The 400 used a four-cylinder version of the engine developed for the 500. It was available in gasoline, LPG, and diesel versions, all of which had 251 cubic inches of displacement and used five main bearings. While the gasoline and LPG tractors were slightly more powerful, all were in the 50-horsepower class, rated for four plow bottoms.

Rather than use the letter designations as before (C for general purpose, O for orchard, etc.), a numerical system was instituted. The first number indicated the size series, the second indicated the engine type, and the third indicated the front end configuration. A 401 was a four-plow/diesel/row-crop tricycle. A 402 was a four-plow/diesel/orchard model. The 410 was a four-plow/gasoline/standard tread. Further variations were added as the need arose.

In a remarkable move, the 400 had an eight-speed transmission—a vast improvement over previous Case tractor's four-speed units. It was the first Case with more than four transmission speeds. Also gone on the new model

Case's Model 300 superseded the Model VAC and was built at the same Rock Island, Illinois, assembly plant. The 300 Model was built from 1955 until 1957, when it was replaced by the 300B Model. Owner: J.R. Gyger, Lebanon, Indiana.

were the hand clutch and chain final drive. The new 400 Series had a proper draft-load compensating three-point hitch and a rubber-torsion suspension seat. Along with the new Tell-Easy instrument panel and the new soft riding seat, it was plain that Case had gotten the message that farmers were interested in comfort.

With the 400 Series, Case also embarked on a new level of advertising. Case employed all media, including television, with glitzy ads emphasizing the new features and paint scheme.

Model 300 and 350

The Model 300 was Case's new three-bottom tractor. Its styling was distinctive and different from the 400, but the color scheme of Desert Sunset and Flambeau Red was the same. Two engines were available: a Case-built gasoline-distillate-LPG engine of 148 cubic inches and a Continental-built diesel of 157 cubic inches. A 4-, 8-, or 12-speed transmission was available. With the 12-speed, the top tractor speed was 20 miles per hour. What do you suppose ex-President Leon Clausen thought of that?

The 300 was built through 1958 with the usual submodel designations that were pioneered with the 400. The 301, for example, was a diesel/row-crop/general purpose. The 350 version came out in 1957. Displacement of the spark-ignition engine was upped to 164 cubic inches, giving it a full 42 horsepower, rather than 36 horsepower of the 300. Also new for the 350 was a 12-volt electrical system.

In 1957, the big 500 diesel was upgraded to the Model 600. Besides getting a new six-speed transmission, the 600 now sported the Desert Sunset paint.

Industrial Division

Case formed an Industrial Division in 1947. The first project was a wheel loader tractor. The hydraulic front end loader was built by Case and factory installed on an SI, DI, VAI, or an

The Case 300 boasted three-range transmissions, duo-control hydraulics, and engines that could be equipped to burn gasoline, LPG, distillate, or diesel. *Case Archives*

LAI. Loaders were called 30, 40, 50, and so on. Previously, other companies made the loaders to fit a variety of tractors. Dealers assembled the loaders, often at a great expense in time for modifications, to make them really fit the tractor.

Case, of course, was experienced in the industrial side of the business from many years back—even back to the steam engine days. The problem was a lack of dealers that would handle and service both the agricultural and industrial customers. Case thought the industrial dealer problem had been solved through an arrangement with Caterpillar of Peoria, Illinois. Cat had no wheel tractors and began selling such Case equipment through its extensive dealer network. Just as the Industrial Division was beginning in 1947, however, Cat pulled out of the arrangement. The reason, they announced, was that they would be marketing a line of wheeled industrial equipment themselves.

Case tried to develop its own dealer network for the industrial side, but the competition was coming on stronger, and Case was losing ground. What the company needed, reasoned the Industrial Division management, was another Caterpillar arrangement—or possibly an acquisition or merger with another crawler manufacturer. They found just what they

High-crop versions of many Case tractor models were offered to farmers in the cane belt. This Model 400 was fitted with an Eagle Hitch. Owner: Jay Foxworthy, Washburn, Illinois.

needed with the American Tractor Company of Churubusco, Indiana.

Marc Rojtman

It was at this point that Marc Rojtman entered the Case picture. Rojtman (pronounced Roitman) was a German Jewish man who left the country seek to refugee from Nazi persecution. Some accounts say he was actually born in Russia and was a refugee from the Bolsheviks as well. A Dunn and Bradstreet report said he had been educated in France. His family had been industrialists, owning a locomotive manufacturing company in France where Marc worked while growing up. At the start of World War II, Rojtman fled to the United States, and it wasn't long before he had established himself as an importer of French movie films. In 1943, Rojtman became a U.S. citizen and enlisted in the Army. With his fluency in English, French, German, and Russian, he was taken into Intelligence Operations. After the war and his discharge from the Service, he worked as a middleman, obtaining contracts for construction equipment from the U.S. government for the European recovery program.

One of the machines handled in this manner was a small crawler, called the Panther, built by the American Steel Dredge Company of Fort Wayne, Indiana. In 1949, Rojtman was able to gain control of American Steel Dredge Company and also of the Federal Machine of Warren, Ohio. Federal, too, was making a small crawler known as the USTRAC. Rojtman's export business had been known as the Washington Tractor Company. Rojtman moved his three operations, which he named the American Tractor Company, to Churubusco, Indiana, near Fort Wayne. From that plant, in 1950, came a line of crawler tractors under the trademark "TerraTrac."

It was rumored that the improvements Rojtman built into the new TerraTracs had been pioneered in the German military Panzer tanks. This was particularly said of the torque converter combined with a hydraulic power-shift transmission. Actually, each track had its own transmission. One could operate forward and the other backward for on-the-spot turning.

American Tractor Corporation (ATC) had soon lined up more than 80 dealers. They specialized in the industrial side of the tractor business, though offering the TerraTracs to farmers as well. In a concession to farmers, Rojtman simply commandeered Ferguson's three-point system. Ferguson was too busy suing Ford over the system to worry about small potatoes like ATC.

Late in 1956, the Case Board of Directors concluded a merger agreement with American Tractor. The agreement included provisions that Marc Rojtman would become executive vice president and general manager of the merged companies. Rojtman had insisted on being named to the presidency, but William Grede, Chairman of the Executive Committee, put his foot down. Rojtman acceded to the lessor role, for a start.

RIGHT
This 1954 ad features the new Model 400 but also lists the VAC-14 and the Model SC.

A 38-year-old human dynamo, Marc Rojtman—with his penchant for fine clothing and his flawless taste in art and furniture—was a composite of P.T. Barnum of circus fame and Bill Lear of aircraft fame. He was talkative and domineering; he had a dynamic leadership style that caused him and his workers to regularly work 15-hour days. A questionable but entertaining story is told of the 5-foot 10-inch, 250-pound Rojtman arriving at Case headquarters for the first time. He and his financial assistant John Grayson roared into Racine in Rojtman's

The Model 400 orchard tractor. Streamlined styling of orchard models often belied the extra engineering involved in getting the operator in a lower position. This often resulted in bizarre linkage for controls and steering. Owner: J.R. Gyger, Lebanon, Indiana.

America's Finest Tractor in the 50 h. p. class...

Two new engines. Powrcel diesel, and Powrdyne for gasoline, LP gas and distillate. Both have big 5-bearing crankshaft, multiple cylinder heads, by-pass controlled cooling, lugging power for hard pulls.

New Powr-Range transmission. Eight speeds forward . . . two reverse . . . with supreme simplicity. Only one shift lever, one clutch, one gear-set. Easy shift pattern.

New power steering makes quick turns in soft ground. Stops road shocks, retains road feel.

New Duo-Valve hydraulic control. Provides down-push to aid penetration, floating action for accurate depth, independent operation of extra portable ram.

Big new Eagle Hitch implements for the "400" give great capacity with all the time-saving convenience of speedy one-minute Eagle Hitch.

CASE

You'll get a thrill when you step onto this totally new tractor, take your choice of eight perfectly-spaced forward speeds, and touch the Duo-Valve hydraulic control. See your Case dealer about it now!

MAIL . . . *for Colorful Folder*

Check here or write in margin any tractor model, any other farm machines that interest you. Address J. I. Case Co., Dept. C-255, Racine, Wis.

☐ "400" Tractor
☐ "500" Diesel
☐ 2-plow "VAC-14" Tractor
☐ 3-plow "SC" Tractor

Are you a student? ..... Acres you farm? ....

Name ....

Address ....

....

CASE

"600" CASE

PREVIOUS PAGE
In 1957, Case introduced the Model 600, which was a makeover of the Model 500 with the additions of a six-speed transmission, different grille, and the line's new Desert Sunset paint on the sheet metal.

white Cadillac El Dorado. Rojtman took one look at the 1904 building, with its mahogany interior, rolltop desks, and brass spittoons. "John," he said, "I can't work in a place like this!"

TerraTrac Line

The line of crawlers from American Tractor prior to the merger were as follows:

The TerraTrac GT-25 was a light (about 3,200 pounds) crawler using a 124-cubic-inch Continental engine of 26 horsepower. A five-roller undercarriage was used. It was the first non-Ferguson to use the hydraulic three-point hitch.

The TerraTrac GT-30 came out in 1951 and was continued through 1954. It was basically the same as the GT-25, except it used the Continental 140-cubic-inch engine. Operating weight of the GT-30 was 4,400 pounds. It had the distinction of being the first tractor to have a maximum drawbar pull greater than its own weight. It was able to pull 4,518 pounds during Nebraska test 471.

The TerraTrac GT-34 was a medium-size crawler using a Continental 162-cubic-inch engine. It was produced from 1951 through 1954. In 1952, a DT-34 diesel version with a 157-cubic-inch Continental engine was added.

For 1955 through the time of the merger, the model designation scheme was changed to a three-number series. These were generally picked up by Case on the other side of the merger.

The first of the new Case crawlers was the Model 310. It was a revision of the American 300, which hailed from the TerraTrac GT-30. The Continental engine was replaced by a Case powerplant of 148 cubic inches, and a general styling was accomplished. A Case diesel of the same displacement was added in 1961.

A big Model 800 came out in 1957. It used a 277-cubic-inch Continental diesel and a torque converter transmission. An updated Model 810 appeared for the 1960 model year. An even larger Model 1010 was also released at that time. It featured a 382-cubic-inch Continental four-cylinder engine. A smaller version, the Model 610, came out at the same time. It used a 208-cubic-inch Continental four-cylinder engine.

Case's first venture into four-wheel drive was this 1200 Traction King Model. The unit featured four-wheel crab steer rather than articulated steering. The Lavona power-cell engine did not respond well to a turbocharger. Owner: John Thierer, Washburn, Illinois.

Case Under Rojtman

The fit between the American Tractor Company and Case was a good one. Marc Rojtman also seemed to be just what Case needed. His aggressive leadership and his company's products represented an exciting complement to Case. Most of ATC's dealers came in on the deal. Case management would not allow them to handle the agricultural lines, however, since they did not want to irritate the dealers they already had.

Rojtman also brought to Case some outstanding ATC personnel. Theodore Haller became vice president of engineering. He had come from Allis-Chalmers to ATC, where he had been chief development engineer for crawlers. Financial wizard John Grayson became Case's controller. David A. Milligan was named sales manager of Case's Industrial Division. Later Rojtman recruited Herbert H. Bloom, former president of Massey-Ferguson's U.S. company, to head foreign operations. Rojtman had the energy and charisma to attract good people, and although he drove them hard, he could generally keep them as well.

The main asset Rojtman brought into the merger was Rojtman himself. He exemplified what one historian of the period called the "new wave" of corporate leadership. The other leaders in the tractor business were also under similar new managers. William A. He-

David Brown Company

The family-owned firm of David Brown had begun business in 1860. By the 1930s, it had grown to be the largest gear producer in Great Britain, with its main plant in Huddersfield, Yorkshire. David Brown, Ltd., was approached by Harry Ferguson in the mid-1930s, and they cooperated in a venture to build a David Brown tractor with a Ferguson implement system. This venture lasted only a few years, before Ferguson left David Brown and entered into an agreement with Henry Ford which led to the Ford—Ferguson 9N. David Brown continued in the tractor business, expanding its line of equipment and tractor size ranges, and growing to a respectable size, when their agricultural division was bought out by J.I. Case Company in 1972. Tractors under 100 horsepower in the Case line were made in England beginning with that purchase.

When Fordson production in the United States ended in 1928, the Irish inventor Harry Ferguson realized there was no likelihood of getting his three-point system incorporated into that tractor. Instead, he began designing a tractor from scratch with the hydraulics built in, not added on. When the drawings were finished, Ferguson began contacting companies specializing in the required parts.

The result was a diminutive Fordson-type machine weighing only a little more than half of the Fordson's weight. The engine was from Hercules, and the transmission and differential were supplied by David Brown, Ltd. The machine was painted black, and because it had no other name it came to be called the "Black Tractor."

Although the Black Tractor had several debilitating shortcomings, it performed well enough to entice David Brown to undertake manufacture of an improved version to be known as the Ferguson-Brown Type A. The first of the Type A tractors was ready for sale in May 1936.

Sales of the Ferguson-Brown were disappointing, as were profits. The tractor could not be made in large enough quantities to bring the costs down, and the price was already too high, especially considering the fact that new implements were needed for the three-point hitch. The tractor itself also had some weaknesses being, in David Brown's eyes, just too small. Brown and Ferguson eventually had a falling out, and Brown, on his own began making the changes he thought necessary.

With that, Harry Ferguson contacted Henry Ford for a demonstration of his tractor. In the fall of 1938, Ferguson and several aides brought a crated Ferguson-Brown tractor and implements by ship and truck to Ford's Fair Lane Estate. Ford had several associates with him there, and all assembled agreed the Ferguson-Brown handily outperformed both a Fordson and an Allis-Chalmers with the same- size plow.

Ford took on a serious attitude and called for a table and chairs to be brought out of the estate's kitchen. Ford and Ferguson seated themselves at the table and Ferguson proceeded to demonstrate the function of the system by using a spring-wound model. The result of this conversation was a handshake, sealing a gentleman's agreement that Ford would build the tractors incorporating Ferguson's Draft Control hydraulic system.

David Brown continued to work on the original concept, but with a more powerful engine and a larger frame. The first David Brown, identified as the VAK 1, came out in 1939. This was improved to become the VAK 1A in 1945, then the Cropmaster in 1947. David Brown entered the diesel era in 1952, adding industrial and crawler models.

A 1946 David Brown VAK 1A Model. First introduced in 1939, David Brown was the original tractor manufactured to use Harry Ferguson's three-point hitch system.

200

Tripl-Range
DRIVE

PREVIOUS PAGE
The 200B was available in either utility (wide) or row-crop front axle configuration. Smallest of the "B" series tractors was the 200B Model. It was so close in size to the next model up, the 300B, that few were made.

RIGHT
A Case 910B Model. Even though this model was made in the late 1950s, it still retained the chain drive system pioneered in 1929. The unit pictured above is powered by LP fuel and was the only configuration of this six-cylinder tractor to use spark ignition.

witt (related to the John Deere family by marriage) was considered a "messiah" at Deere. Archie Cardell had taken over at International Harvester. And Ford Motor Company was now being run by the Whiz Kids. Rojtman burst on the Case scene a consummate huckster and wheeler-dealer of enormous energy and ego.

Frank Palermo, later vice president of manufacturing and no fan of Rojtman, defended Rojtman's tenure at Case. "He may not have reached all his goals, but in trying, he took the company to new heights. He painted his picture of the future in broad strokes, backed by astounding optimism and energy."

Rojtman left his Milwaukee Lake Shore Drive "chateau" late each morning to drive to his Racine office. There he would usually remain until ten at night, or much later. He would often call subordinates in the middle of the night—and when he spoke, people jumped!

Rojtman was, above all, a promoter. The Board resisted his demands to increase production because there was no market. Rojtman's response was "create a market." To prove his point, he staged a "World Premier in Phoenix."

For the first time ever, an agricultural equipment manufacturer had, with much fanfare, gotten their entire dealer network together for the unveiling of a new product line. The extravaganza last-

CASE
900

ed for six weeks. Chartered planes flew dealers, their spouses, their bankers, and anyone else Rojtman could think of for three-day visits to Phoenix. They were wined and dined and entertained by big-name entertainers. It was an awesome experience for most. To be whisked from the frigid Midwest to warm and sunny Phoenix and treated like royalty was overwhelming.

The dealers responded. They ordered 30,000 tractors alone. The Premier had cost $1 million, but that represented only 1 percent of the sales generated. The success of the event was not lost on the competition. In 1960, when Deere was retiring its famous two-cylinder tractor line in favor of modern multi-cylinder units, Deere's William Hewitt staged the "Deere Days in Dallas," renting the Coliseum and flying in dealers from all over.

With no regard for tradition in nomenclature, Rojtman ordered the next generation of tractors (to come out in 1957) to have a new and unrelated numbering system. This new numbering was called the B System, because a "B" was added to the designator number. This was done to differentiate the new tractor numbers. The 600 became the 900; the 400 became the 700 and 800; and the 300 and 350 became the 200, 300, 400, 500, and 600. Needless to say, there was some confusion. The Desert Sunset and Flambeau Red paint scheme remained the same, but the headlights were now placed in the grille. New styling followed that of other industrial items and even the automobile. The hood and grille lines were thrust forward and squared off. New grilles were practical and purposeful. Chrome

By the late 1950s, Case was making mist of their smaller models available in utility styles as well. These units were built for turf or warehouse work. Pictured here is a 430 utility Model. Owner: J.R. Gyger, Lebanon, Indiana.

By the early 1960s, Case had strong involvement with a dealership in Winter Garden, Florida, the Pounds Motor Company. So strong was this dealership's position with Case that "Florida Special" models were made to their specifications. Pictured above is a Model 630 diesel with Case-O-Matic and a hand clutch. Owner: J.R. Gyger, Lebanon, Indiana.

stripes and letters replaced decals.

The big news was the Case-O-Matic torque converter drive. It gave infinitely variable speeds and doubled pulling power. It was available on the Model 800 and larger tractors, except for the 900. By 1958, Case was offering 12 power sizes of tractors, for a total of 124 distinct models. This included the crawlers mentioned previously.

New Case Hundreds

The Model 900 retained the hand clutch and chain drive of its heritage dating back to the Model L of 1929. The engine originated in the original 500 Diesel. The engine was now capable of 71 horsepower because, at Rojtman's insistence, rated engine speeds were upped to their maximums, in this case 1,500 rpm. The six-speed transmission was retained, and the Case-O-Matic torque converter was not available. An LP-gas version was optional using the same basic engine. The 900B was built in 1957 and 1958 only in the standard tread version.

With the exceptions of the new styling and the Case-O-Matic drive, the 800B was essentially the same as the 400. An increase in rated engine speed gave it more power, however. The 700 was the same as the 800, but with a standard transmission rather than Case-O-Matic and the rpm increase.

The 600B harked back to the Model 350 of pre-1957 days. The tractor featured an eight-speed transmis-

NEXT PAGE
Aerial view of the Case world premiere new model introduction, held in Phoenix, Arizona, in 1960. The gala event cost the company over a million dollars but helped put the firm on a stronger financial footing.

WORLD PREMIERE

TEST TRACK
1960 Case-o-matic Line

CAS
Dual-Range DRIVE
CASE

LEFT
The Case 730 series replaced the 700 series in 1969. Available in three fuel configurations, the 730 was one of a now-large line of size ranges in the Case system.

sion plus the torque converter. The torque converter could be locked out in each of the regular gears, or released, for a total of 16 speeds. The 500B was identical with a standard transmission. The 300B was the same as the old 300. The new 400B was the same as the old 300C and sported the Case-O-Matic drive.

The 200B was smaller than the old 300, but with a 127-cubic-inch four-cylinder gasoline-only engine. It was available in row-crop and utility versions.

Loader Backhoe

An American Tractor Company dealer named Walter Smith came to Churubusco at the time of the Case-ATC merger to help ATC with its display for the visiting Case people, dealers, and the public. Case had brought some of their equipment to Churubusco as well. After hours, a group of ATC people, including Smith, were looking over a Case 300 tractor with a front end loader. One of them commented that the ATC backhoe (which ATC had been building for crawlers) might fit on the 300 wheel tractor. They rolled the 300 into the shop and, as Smith said, "whomped up" a 300 Loader-backhoe.

Rojtman was furious when he saw it. So were the Case executives. Nevertheless, the unit was shown to the public with the other equipment. It drew rave reviews, more-or-less vindicating Smith and the ATC people involved.

Next, two ATC engineers, Paul Hawkins and Elton Long, were assigned to do a proper design. They proceeded

The unit pictured above is a high-crop Model 830 HC (high-clearance) tractor, in Comfort King configuration, with Case-O-Matic transmission. Owner: J.R. Gyger, Lebanon, Indiana.

1030
CASE

Late production 30 series tractors had Draft-O-Matic, Case's own link sensing draft control. Earlier 30 series units still used the Eagle Hitch system. Owner: John Thierer, Washburn, Illinois.

with the 300 Agricultural tractor, adding a torque converter, Case's loader, and ATC's backhoe. They added "beef" as necessary. Later, a power-shift shuttle control was added. Production was in the Burlington plant. Its introduction marked the dawning of a new age for Case and, especially, the Industrial Division.

Thirty Series

Beginning with the model year 1960, each tractor model number consisted of three numbers, the last two of which were "30." Thus, the 600 became the 630, the 700 became the 730, and so on. The tractors were essentially the same, but with transmission options increasing exponentially. There were now triple-range transmissions, shuttle-shift transmissions, and hand and foot clutches used together. There were also some new engine displacements, but the same four or five blocks were used. Factory cabs were also offered.

In 1965, the CK, or Comfort King series, replaced the previous line. For the first time, the 900CK was available in a gasoline version and as a row crop.

For 1966, a Model 1030 was added to the lineup. It was available as a general purpose model, or as a Western (standard tread) model. Featured was a new 451-cubic-inch diesel rated at 2,000 rpm. It was Case's first general purpose tractor with over 100 horsepower.

The same 451-cubic-inch engine was used in Case's first four-wheel-drive tractor, the 1200 Traction King. This 17,000-pound monster featured four-wheel steering.

Exit Rojtman

Before the ink was dry on the merger papers of 1956, Marc Rojtman had made one powerful enemy on the

LEFT
During the early 1960s, most major tractor makers rushed to put their names on garden tractors. Case contracted with the Colt Manufacturing Company of Winnecone, Wisconsin, to make these machines. Owner: Weston Link, Ridgeland, Wisconsin.

Case Board: Leon Clausen. It was to be expected that some friction would erupt between such dramatically different forceful men, but not so soon. One of Rojtman's first suggestions had been to modernize the trademark eagle Old Abe. Clausen was fit to be tied and, thereafter, opposed Rojtman at every turn. When the Board made Rojtman president, Clausen resigned from the Board.

William Grede, Clausen's heir apparent, and other conservatives on the Board were also uneasy with Rojtman's overseas expansion policy. Indeed, Rojtman here, as usual, was overrunning his own headlights and some of these ventures went sour. They did, nevertheless, put Case in a good international posture by the end of the 1960s, when it really counted.

Rojtman abhorred the Board's conservatism. He viewed their policy as that of doing nothing, thereby doing nothing wrong. His aggressiveness appeared to be paying off, as sales were up dramatically for 1957. The Board members worried that this was only wholesale; dealer inventory was increasing to an unacceptable level. Rojtman's scheme was predicated on continued expansion, but after 1968, there was an economic downturn that hit the overstocked dealers in a bad way. Case was left with too much in Accounts Receivable and with excessive debt. While the Board panicked, Rojtman begged them to stay the course and ride out the downturn, which he predicted would be short. Instead, the Board ordered that he cut production.

RIGHT
Case garden tractors were built by the Colt Manufacturing Company. The firm was eventually purchased by Case and is now the Ingersol Company.

it's a family affair
Drive a Case 10 or 12 horsepower tractor and you'll know why CASE calls them "the compacts." They're ideal for so many of the farm jobs too big to do by hand . . . too small for bigger tractors. And so rugged and powerful! Won't bog down when the going gets tough. The perfect size tractor to clean cramped pens, poultry houses and barns . . . grass and weed cutting . . . tilling gardens . . . snow removal. Great as a portable source of mechanical or hydraulic power to run elevators, generators, compressors, pumps, mixers and hoists. **You'll enjoy driving** a CASE compact. No clutch, no shift Hydra-Static Drive . . . large flotation tires . . . push button starting . . . comfort foam padded seat . . . and many other features as standard equipment. Designed and built by CASE which makes all sizes of farm tractors. Write for your free copy of the new Case Compact Tractor booklet, J. I. Case Company, Dept. 103, Racine, Wisconsin 53404
CASE®
great for cramped areas
hydraulic roto-tiller
portable electric power
mechanical pto

A 1030 diesel Comfort King unit shown here with a cab option. This was Case's largest tractor in the 30 series, with over 100 horsepower. Owner: Jay Foxworthy, Washburn, Illinois.

At about this time, Rojtman held another of his several dealer extravaganzas, this one in Nassau. Falling victim to his own hoopla, he defied the Board and increased production.

Previously when Rojtman and the Board locked horns, Rojtman made it a practice to offer to resign in order to get his way. This he did when confronted about increasing production. This time the Board's response was different: They accepted the resignation on the spot. The resignation offer had been anticipated. The Board had William Grede in place to replace Rojtman.

Thus on February 1, 1960, after less than four years, Marc Rojtman was out. In that time he had transformed a doddering giant into a worthy competitor to number one Deere & Company. It was now up to 63-year-old William Grede to lead.

William Grede was from the same cloth bolt as Clausen. He was an extreme political and social conservative and one of the founding members of the John Birch Society. He was not a college man but had invested in a small, but successful foundry in Milwaukee that became Grede Foundries. He was invited to join the Case Board in the first place because he had been able to keep unions out of his factories. Ironically, Grede's policy toward Case's bargaining unit led to a six-month strike. During the strike, much of Case's inventory and cash flow problems were solved.

This was still not enough for the bank creditors. Over the years 1960 and 1961, Assistant Controller James Ketelsen and other executives struggled with options and plans that would allow Case the necessary financing to continue operation. Finally, an agree-

In the mid-1950s, Case bought the small American Tractor Company, a firm that built crawlers and hoes. The unit pictured above is an early 320 loader-hoe combination built shortly after the ATC purchase. *Robert Pripps*

ment was reached. One of the stipulations, however, was that Grede had to go as president of Case. The difference between Grede and Rojtman was so striking that Harvard Business School used them as a management case study. Each contributed to the survival of Case, but neither "survived" to participate in the fruits of their labor.

Infusions of Cash

Needless to say, in 1961 Case was in a precarious financial position—and now it was "leaderless." By March 1962, however, a new, strong leader was found in the person of Merritt D. Hill, who had been president of Ford's Tractor and Implement Division. With a strong cast of supporting characters, some good products, and some well-timed acquisitions, Case seemingly had weathered another storm. One of the acquisitions was the Colt Manufacturing Company of Winneconne, Wisconsin. Colt made small lawn and garden tractors for the homeowner market; a market that was proving to be a good one for Deere & Company.

Case was still failing to generate enough cash to fuel growth and still pay its bills. Clearly, a financial partner was needed. An arrangement was made with the Kern County Land Company (KCL), a California investment company dating back to the 1848 Gold Rush. KCL sought tax advantages by acquiring controlling interest in Case. KCL's backing allowed Case to expand dramatically between 1964 and 1967.

In 1966, Merritt Hill became Board Chairman, and Charles Anderson, who came to Case from KCL, took over as president. It wasn't long before KCL had problems of its own: a hostile takeover by buccaneer Armand Hammer and Occidental Petroleum. KCL fled into the arms of "white knight" Tenneco. The deal was closed in August 1967. Tenneco acquired, and placed under Case, construction equipment manufacturing companies Drott,

Universal, and Davis. Also acquired was Beloit Woodlands, a timber harvesting equipment company. Tenneco also replaced Charles Anderson as president with James Ketelsen, the financial wizard who had saved the company during the Rojtman-Grede era. The years 1968 and 1969 marked another recession in the agricultural industry. President Ketelsen responded by "aiming for minimum participation in the implement market." This was his euphemism for becoming a tractor-only company.

70 Series

For some time, Case had been preparing a new line of tractors to replace the Comfort King series. The new line, the 70 Series, would be called the Agri-Kings.

The smallest of these was the 470, which came out in 1969. It was available in general purpose, standard tread, and low-profile models in both diesel and gasoline versions. A similar 570 was also offered.

The 770 also came out in 1969, but as a 1970 model. The 770 featured an entirely new power-shift transmission, although a manual was still available. It was available as a diesel only.

The 870 and 970, on the other hand, were sold in gasoline versions through 1973, and as diesel-only through 1975 (1978 for the 970). Heated and air conditioned cabs became a popular option during this period.

The 1070 was Case's 100-plus horsepower tractor. It featured a Category II three-point hitch with lower link draft control sensing. This was the approach Ferguson initially had tried,

The 320 Case-loader-backhoe was the first in the industry to be manufactured, sold, and warranteed by a single company. *Case Archives*

CASE

The Model 470 was only produced for a few months. This was due to the purchase of the British David Brown, Ltd., firm shortly into the 70 series production. From that time on, Case's small tractors were produced in England, and the smaller 70 series tractor production was diverted to Case's Burlington, Iowa, backhoe plant. Owner: Weston Link, Ridgeland, Wisconsin.

but without the aid of electronics he could not get it to work. The 1170 model was the same as the 1070, except the engine was turbocharged. There was also an 1175 model that was essentially the same, except it featured the new roll-over-protection-system (ROPS). The 1270 was the same configuration and used the same engine, but the engine was rated for slightly more horsepower. The 1270 was sold from 1972 through 1978.

The 1470 was a Traction King four-wheel drive, four-wheel crab-steer tractor, the largest Case had built. It came out in 1969. It featured Case's new 504-cubic-inch direct injection diesel developing 145 horsepower. It was sold through 1972.

The famous 1570 turbocharged diesel was the largest two-wheel-drive tractor in its time frame: 1976–1978. The 504-cubic-inch six-cylinder engine produced 180 horsepower at 2,100 rpm.

Introduced in 1971 and 1974 respectively, the big Case 2470 and 2670 four-wheel-drive tractors featured rigid-frame four-wheel steering. The 2670 was Case's most powerful tractor up to that time at 219 horsepower.

For the 1976 to 1978 period, Case brought out the Model 1570. It was a turbocharged diesel with a 12-speed transmission. A cab was standard. The engine was a 504-cubic-inch six-cylinder unit capable of 180 horsepower. During the first half of the 1976 model year, 1570s were offered with a special red-white-and-blue stars and stripes paint job to commemorate the country's bicentennial.

The 580 Super K construction King Loader-backhoe is the current extension of the original Model 320 made in the mid-1950s. *Case Archives*

In 1969, Case introduced this new logo to replace Old Abe, the original company trademark that harked back to the Civil War. Designed by the New York firm of Lippincott and Margolis, the new logo represents the Case name in metal treads to better reflect the company's emphasis on construction equipment.

The End for Old Abe

The grand old man of Case, Leon Clausen, died in 1965. Clausen had risen in anger when Marc Rojtman had proposed modernizing the trademark eagle, Old Abe, in 1956. From then until Clausen died, no one dared raise the issue again. Old Abe had perched atop the globe symbolizing the Case Corporation for more than 100 years. Now without the fear of Clausen's retribution, Morris Reed (executive vice president) retained the firm of Lippincott and Margulies of New York to redesign the Case logo. One of the reasons for the retirement of Old Abe was that the company was marketing aggressively to the construction industry, where the eagle trademark was not well recognized. The new trademark consisted of Case's name "stamped in metal treads" of power red, black, or white.

Case was not concentrating entirely on the construction business, however. In 1972, the company bought David Brown, Ltd., a British agricultural equipment firm. David Brown had manufacturing plants in Leigh and Meltham in the United Kingdom.

When Case purchased Brown, they acquired much-needed agricultural outlets in Europe. The Case 470 and 570 tractors were discontinued at Racine. Racine's Clausen plant, named for the former president, produced only larger tractors. Tractors smaller than the 770 were made in Great Britain at the David Brown facilities.

In recognition of the Case-David Brown marriage, agricultural tractors were painted in a new color scheme: David Brown White sheet metal, with Case Power Red castings.

Case's Black Knight was another dealer promotion for the 870 series, built between 1970 and 1975. Owner: J.R. Gyger, Lebanon, Indiana.

The 1370 series was built beginning in 1969. The term "Agri-King" was first used with the 70 series. Powered by Case's own 504-cubic-inch diesel, it was the first series to feature a standard power-shift transmission.

The Model 1570 built in the 1970s was offered for a short time during the 1976 centennial year as a dealer promotion model in this "Spirit of 76" model. Owner: J.R. Gyger, Lebanon, Indiana.

AGRI KING
GYGER
SPIRIT OF '76
case

Appendix

Clubs, Shows, and Literature

Clubs, Newsletters, and Magazines

Magazines and newsletters that provide a wealth of information and lore about individual brands of antique farm tractors and equipment have been on the scene for some time. More are springing up each year, so the following list is far from complete. Many of these publications come with collector club memberships.

Case Sources

Old Abe News (Case Collectors Assoc., Inc.)
Dave Erb, editor
4004 Coal Valley
Vinton, OH 45686-9741

Tractor-Related Clubs and Publications

Antique Power
Patrick Ertel, editor
PO Box 838
Yellow Springs, OH 45387

Belt Pulley (Antique Tractors)
Kurt Aumann, editor
PO Box 83
Nokomis, IL 62075

Engineers and Engines
Donald Knowles, editor
1118 N. Raynor Avenue
Joliet, IL 60435

Farm Antique News (Tractors and Antiques)
Gary Van Hoozer, editor
PO Box 96
Tarkio, MO 64491

Gas Engine Magazine
Linda Sharron, editor
PO Box 328
Lancaster, PA 17603

Iron Men Album
Gerald Lestz, editor
PO Box 328
Steam Lancaster, PA 17603

The Tractor Magazine (Antique Tractors)
Steve Sharp, editor
PO Box 174
Spencer, NE 68777

Other Specialty Clubs and Publications

Antique Caterpillar Machinery Owners Club
Marv Fery, editor
10816 Monitor-McKee Road NE
Woodburn, OR 97071

Ferguson Club Journal
Ken Goodwin, editor
Denehurst, Rosehill Road
Market Drapton
TF9 2JU England

Fordson Club News
Tom Brent, editor
Box 150
Dewdney, B.C.
VOM 1HO Canada

Golden Arrow (Cockshutt and Co-op)
John Kasmiski, editor
N7209 State Hwy. 67
Mayville, WI 53050

Green Magazine (John Deere)
R. and C. Hain, editors
RR 1
Bee, NE 68314

Hart-Parr/Oliver (Collector)
Kurt Aumann, editor
PO Box 687
Charles City, IA 50616

M-M Corresponder (Minneapolis-Moline)
Roger Mohr, editor
Rt 1, Box 153
Vail, IA 51465

9N-2N-8N Newsletter (Ford)
G.W. Rinaldi, editor
P.O. Box 235
Chelsea, VT 05038-235

Old Allis News (Allis-Chalmers)
Nan Jones, editor
10925 Love Road
Belleview, MI 49021

Oliver Collector's News
Dennis Gerszewski, editor
Rt 1
Manvel, ND 58256-0044

Plug 'N Points (Antique Trucks)
Tom Brownell, editor
RR 14, Box 468
Jonesboro, TN 37659

Prairie Gold Rush (Minneapolis-Moline)
R. Baumgartner, editor
Rt 1
Walnut, IL 61376

Red Power (International Harvester)
Daryl Miller, editor
Box 277
Battle Creek, IA 51006

Two-Cylinder Club (John Deere)
Jack Cherry, editor
PO Box 219
Grundy Center, IA 50638

Wild Harvest (Massey-Harris, Ferguson)
Keith Oltrogge, editor
1010 South Powell, Box 529
Denver, IA 50622

Tractor Shows

For an annual directory of engine and threshing shows, contact Stemgas Publishing Company, P.O. Box 328, Lancaster, PA 17603; (717) 392-0733.

The cost of the directory has been $6.00. It lists shows in virtually every area of the country. Stemgas also publishes *Gas Engine* and *Iron Men Album*, magazines for the enthusiast.

Sources and Recommended Reading

The following books offered essential background on the origins and history of J.I. Case and the tractors and equipment of the times. These make good reading and good library additions for any tractor buff. Most are available from Motorbooks International, P.O. Box 1, 729 Prospect Avenue, Osceola, WI 54020; (800) 826-6600.

The Agricultural Tractor 1855–1950 by R.B. Gray, Society of Agricultural Engineers. An outstanding and complete photo history of the origin and development of the tractor.

The American Farm Tractor by Randy Leffingwell, Motorbooks International. A full-color hardback history of all the great American tractor makes.

The Century of the Reaper by Cyrus McCormick, Houghton Mifflin Company. A firsthand account of the Harvester and Tractor Wars by the grandson of the inventor.

Case Tractors by Andrew Morland and Nick Baldwin, Motorbooks International. A colorful look at Case tractors.

Case: A Photographic History by April Halberstadt, Motorbooks International. A black-and-white look at the Case corporation.

Classic American Farm Tractors by Andrew Morland and Nick Baldwin, Osprey. A superb color documentary of some of the great old tractors.

Encyclopedia of American Farm Tractors by C.H. Wendel, Crestline Publishing. Notes and data on all the old (and some obscure) tractors.

Farm Tractors 1926–1956 by Randy Stephens, editor, Intertec Publishing. A compilation of pages from the *Cooperative Tractor Catalog* and *Red Tractor Book*.

Farm Tractors 1950–1975 by Lester Larsen, The American Society of Agricultural Engineers.

Fordson, Farmall and Poppin' Johnny by Robert C. Williams, University of Illinois Press. A history of the farm tractor and its impact on America.

Ford Tractors by Robert N. Pripps and Andrew Morland, Motorbooks International. A full-color history of the Fordson, Ford-Ferguson, Ferguson, and Ford tractors, covering the influence these historic tractors had on the art of tractor design.

Full Steam Ahead, Volume 1: J.I. Case Tractors and Equipment 1842–1955, by David Erb and Eldon Brambraugh, American Society of Agricultural Engineers (ASAE). This comprehensive volume covers the development of the Case tractors and the company.

Great Tractors by Michael Williams and Andrew Morland, Blandford Press. Describes, in words and pictures, tractors that were milestones in history.

Harvest Triumphant by Merrill Denison, WM. Collins Sons & Company. The story of human achievement in the development of agricultural tools (especially in Canada), and the rise to prominence of Massey-Harris-Ferguson (now known as the Verity Corporation). Rich in the romance of farm life in the last century and covering the early days of the Industrial Revolution.

Henry Ford and Grass-roots America by Reynold M. Wik, The University of Michigan Press. Recounts the era of the Tin Lizzie and the Fordson.

How Johnny Popper Replaced the Horse by Donald S. Huber and Ralph C. Hughes, Deere & Company. The history of the two-cylinder tractor from the perspective of Deere & Company.

How to Restore Your Farm Tractor by Robert N. Pripps, Motorbooks International. Follows two tractors through professional restoration. Includes tips and techniques, commentary, and photos.

John Deere's Company by Wayne G. Broehl, Jr., Doubleday and Company. A scholarly tome on the history of Deere & Company and its times.

Nebraska Tractor Tests Since 1920 C.H. Wendell, Crestline Publishing. Consolidated descriptions of all the tractor tests conducted by the University of Nebraska. A must for any old-tractor enthusiast!

150 Years of J.I. Case by C.H. Wendel, Crestline Publishing. A complete photo-documented product history.

Traction Engines by Andrew Morland, Osprey. Descriptions and color photos of most of the great old steam tractors.

Index

A-6 combine, 59
Agitator thresher, 13, 15
Agri-King, 124
Aircraft engine, 29
All-gray tractors, 35
American Tractor Company, 111, 118
ATC backhoe, 111
Automobile line, 15, 25, 30

Bear, Wallis, 17
Big 4, 18
Big Four, 14, 16
Bitter battle, 31
Black Knight, 123
Brantingham, Charles, 18
Brown, David, 77
Brown, John T., 93
Bull Tractor Company, 27
Bull, Frank, 27
Bull, Stephen, 16

Car business, 30
Case engine (150 horsepower), 26
Case garden tractors, 114
Case Low-Seater, 77
Case steam engine, 11
Case thresher, 32, 36
Case V models, 81
Case VAO, 88
Case VC models, 81
Case's 110-horsepower traction engine, 26
Case, Caleb, 8
Case, Jackson I., 16
Case, Jerome Increase, 8, 15, 19
Case-O-Matic, 107
Case-Paterson tractor, 15
CC4, 49, 51, 52
CH, 55
Chicken roost, 54
Chicken-roost steering, 83
Chief Sky, 14
Clausen, Leon R., 40, 41, 44, 61, 70, 123
Colt Manufacturing Company, 114
Continental four-cylinder engine, 70
Crossmotor tractors, 9, 35
Crossmotor, 30, 36, 40
Crozier, Wilmot F., 33
Cub J, 29
Cutting grain, 13

D series, 66, 71
David Brown Company, 101
Davis, John, 9
Depression, 49
Draft-O-Matic, 113

Eagle Hitch system, 73
Eagle Hitch, 83
Eclipse, 11, 15
Ela, Richard, 11
Emerson, Ralph, 17
Emerson-Brantingham Company, 17
Erie Canal, 9

Ferguson, Harry, 72, 76, 77
First Case steam engine, 24
First Case tractor, 14
First self-steering Case steam traction engine, 24
Flambeau Red, 65
Florida Special, 107
Ford, Henry, 29, 72, 81
Fordson, 17, 29, 41, 48, 58
Froelich, John, 24

Garden tractors, 114
General Motors, 17

Haller, Theodore, 100
Hart-Parr, 24
Heider, Henry, 19
Heider, John, 19
Hill, Merritt D., 118
Horse power over mechanical power, 22

Industrial Division, 94

J.I. Case Plow Works, 16
Jackson, Andrew, 8

Little Bull, 27
Live PTO, 72
Loader Backhoe, 111
LP gas, 107
Lubricator, 32

Manny, Pells, 17
Massey-Harris, 29
McCann, Dan, 14
Milligan, David A., 100

Models

- 9-18, 31, 36
- 9-18B, 31, 32
- 10-18, 34
- 10-20, 29-31
- 12-20 tractor, 36
- 12-20, 33, 37
- 12-25, 27
- 15-27, 9, 29, 36
- 15-27, 37
- 18-32, 38
- 20-40, 22-24, 27
- 22-40, 30, 32, 36
- 22-40, 37
- 25-45, 33
- 30-60, 27
- 30 series, 113
- 30 series, 113
- 40-72, 38
- 70 series, 119
- 200B, 104
- 300, 94
- 300, 95
- 320 loader-backhoe, 118, 119
- 400 orchard tractor, 96
- 400, 93
- 430 utility Model, 106
- 470, 121
- 500, 73, 76, 86, 93
- 580 Super K construction King, 122
- 600, 94
- 630 diesel, 107
- 730, 111
- 830 HC, 111
- 900, 107
- 910B Model, 104
- 1030 diesel Comfort King, 117
- 1200 Traction King, 100
- 1270, 121
- 1370, 124
- 1570, 121, 124
- A-6 combine, 45
- Agitator threshing machines, 13
- automobile, 30
- Big 4, 18
- C with Trackson conversion, 53
- C, 43, 47, 48, 54
- CC, 44, 45, 48, 49, 51-53, 59
- CI, 52
- D, 84
- DO, 86
- Eclipse thresher, 11
- Garden tractors, 114
- Hundreds, 1007
- K, 29
- L engine, 55
- L, 18, 44, 46-48, 58-60
- LA, 84, 91
- Low-Seater, 77
- Motor cultivator, 27, 44
- OK, 29
- R, 61-63
- RC, 45, 46, 53, 59, 63
- RC, 62
- S series, 65-67, 78
- S, 67, 88
- SC model , 65, 67
- SI, 91
- Steam engine, 11
- Steam engines, 11, 26
- Thresher, 32, 36
- Thresher, 40x58, 24
- V, 81
- VA series, 69, 70, 81, 82, 86
- VAC, 82
- VAC-14, 77
- VAH, 86
- VAI, 89
- VAO, 88
- VC models, 81

Motor cultivator, 27, 44

Nebraska Tractor Test Law, 31, 32
New logo, 122

Old Abe, 8, 14, 15, 17, 123
Orchard versions, 86
Oswego, New York, 8
Otto cycle (four-cycle) engine, 24

Palermo, Frank, 104
Paterson's patent drawing, 15
Paterson, William, 15
Pierce Motor Company, 30
Pitts Brothers, 8
Pitts Groundhog threshers, 8, 9
Pope threshing machine, 8
Production line, 40
PTO-driven binder, 78

Racine County, Wisconsin, 9
Roadless and Trackson, 53
Rock Island Model V, 81
Rock Island Plow Company, 18
Rockford, Illinois, plant, 18
Rojtman, Mark B., 93, 96, 104, 113
Roll-over-protection system (ROPS), 121
Rubber tires, 52

Samson, 17
Smith, Walter, 111
Steam engines, 41
Steam power shows, 78
Steam traction machines, 21
Steam, 23
Sunburst grille, 54

Talcott, Waite, 17
Tate, Robert N., 18
Tenneco, 11, 118
TerraTrac Line, 100
Three-Point Hitch, 72
Threshing scene, 21
Torque converter, 107
Traction King, 121

Wallis Cub Junior, 29
Wallis, H.M., 16
Warner, Seth, 11
Waterloo Gasoline Traction Engine Company, 24
Watt, James, 23
Waukesha engines, 62
Whiting, Ebenezer, 16
William Deering Company, 25
World War II Production, 80